210
Advances in Anatomy, Embryology and Cell Biology

Co-ordinating Editor

H.-W. Korf, Frankfurt

Editors

H.-W. Korf • F.F. Beck • F. Clascá • M. Frotscher
D.E. Haines • N. Hirokawa • Z. Kmiec • R. Putz
J.-P. Timmermans

For further volumes:
http://www.Springer.com/series/102

Doychin N. Angelov

Physical Rehabilitation of Paralysed Facial Muscles: Functional and Morphological Correlates

With 22 figures

Springer

Prof. Dr. Doychin N. Angelov
Institut I für Anatomie der Universität
Joseph-Stelzmann-Str. 9
50931 Köln
Germany
angelov.anatomie@uni-koeln.de

ISSN 0301-5556
ISBN 978-3-642-18119-1 e-ISBN 978-3-642-18120-7
DOI 10.1007/978-3-642-18120-7
Springer Heidelberg Dordrecht London New York

Library of Congress Control Number: 2011922244

Cover design: deblik, Berlin, Germany

Cover figure kindly provided by Corina Schmidt, Ph.D., University of Ulm, Germany

Printed on acid-free paper

Springer is part of Springer Science+Business Media (www.springer.com)

In memory of my long-standing friend and neuroanatomy teacher Prof. Dr. Kamen G. Usunoff (Department of Anatomy, Medical University Sofia, Bulgaria) who died suddenly on March 1, 2009 during a scientific stay in Rostock, Germany. His sound knowledge, catching enthusiasm, and endless energy for neuroscience research inspired young scientists worldwide for decades.

Abstract

Using a combined morphofunctional approach, we recently found that polyinnervation of the neuromuscular junction (NMJ) is the critical factor for recovery of function after transection and suture of the facial nerve. Since polyinnervation is activity-dependent and can be manipulated, we tried to design a clinically feasible therapy by electrical stimulation or by soft tissue massage. First, electrical stimulation was applied to the transected facial nerve or to paralyzed facial muscles. Both procedures did not improve vibrissal motor performance (video-based motion analysis of whisking), failed to diminish polyinnervation, and even reduced the number of innervated NMJ to one-fifth of normal values. In contrast, gentle stroking of the paralyzed vibrissal muscles by hand resulted in full recovery of whisking. Manual stimulation depended on the intact sensory supply of the denervated muscle targets and was also effective after hypoglossal–facial anastomosis, after interpositional nerve grafting, when applied to the orbicularis oculi muscle and after transection and suture of the hypoglossal nerve. From these results, we conclude that manual stimulation is a noninvasive procedure with immediate potential for clinical rehabilitation following facial nerve reconstruction.

Acknowledgment

This work has been supported by the Jean Uhrmacher-Foundation, the Imhoff-Foundation, and the Köln Fortune Program.

Special thanks to my colleagues and friends Prof. Dr. Athanasia Alvanou, Prof. Dr. Sarah Dunlop, Dr. Emilia Evgenieva, Dr. Maria Grosheva, Dr. Marcin Ceynowa, Prof. Dr. Orlando Guntinas-Lichius, Dr. Gregor Hundeshagen, Privatdozent Dr. Andrey Irintchev, Assoc. -Prof. Katerina Kaidoglou, Dr. Thomas Paling, Dr. Stoyan Pavlov, Privatdozent Dr. Nektarios Sinis, Dr. Emmanouil Skouras. The skillful technical assistance of Diana Bösel, Kathrin Glück, Dirkje Felder, Nadin Lange, Jürgen Rahn, Madlenn Strauss, Lena Wilken, and Claudia Zynthek is highly appreciated.

Contents

Chapter 1
Factors Limiting Motor Recovery After Facial Nerve Injury

The facial nerve is the most frequently damaged nerve in head and neck traumata. Apart from traffic-accident injuries (temporal bone fractures or lacerations of the face), most facial nerve lesions are postoperative (removal of cerebellopontine-angle tumors, parotid resections because of malignancy). Despite the use of fine microsurgical techniques for repair of interrupted nerves in man, the recovery of voluntary movements of all 42 facial muscles and emotional expression of the face remain poor (Vaughan and Richardson 1993; Ferreira et al. 1994; Anonsen et al. 1986; Goodmurphy and Ovalle 1999), and the occurrence of a "postparalytic syndrome" (pareses, abnormally associated movements, and altered reflexes) is inevitable (Kimura et al. 1975; Bento and Miniti 1993; Kerrebijn and Freeman 1998). This insufficient recovery has been attributed to:

1. Persisting posttraumatic rearrangements of motor cortical representation areas (Sanes et al. 1990; Franchi 2000; Taylor et al. 2009). This, in turn, is associated with an acute deafferentation of the axotomy-lesioned motor nuclei ("synaptic stripping", Blinzinger and Kreutzberg 1968) which causes alterations in the synaptic input to the hyperexcitable facial motoneurons (Graeber et al. 1993; Moran and Neely 1996; Brännström and Kellerth 1999)
2. Extensive collateral branching of axons at the lesion site causing "misdirected" or "aberrant" reinnervation of the targets (Montserrat and Benito 1988)
3. Exchange of nerve impulses between adjacent axons (Sadjadpour 1975)
4. Vigorous intramuscular sprouting of axons in the facial muscles

1.1 Altered Synaptic Input to the Axotomized Hyperexcitable Facial Motoneurons

After transection of the facial nerve, the resident microglia show a dramatic increase in mitotic activity, rapidly migrate toward the neuronal cell surface (Rotter et al. 1979), and displace the afferent axo-somatic synaptic terminals (Blinzinger and Kreutzberg 1968). This "synaptic stripping" leads to a deafferentation mainly of proximal but not of peripheral dendrites (Bratzlavsky and vander Eecken 1977; Titmus and Faber 1990; Nacimiento et al. 1992). The axotomized motoneurons "respond" to their deafferentation with a decrease in the synthesis of transmitter-related compounds, for example, muscarinic and glycine receptors (Senba et al. 1990) and a decrease in activity of enzymes involved in the biosynthesis of transmitters, for example, dopamine-β-hydroxylase, tyrosine-hydroxylase, cholineacetyltransferase, cytochromeoxidase, and acetylcholinesterase (Engel and Kreutzberg 1986; Engel et al. 1988). These changes correspond to the electrophysiological status of regenerating neurons: increased excitability (Eccles et al. 1958; Kuno and Llinas 1970) with preserved integrity of the dendritic input (Lux and Schubert 1975; Kreutzberg et al. 1975; Borgens 1988).

1.2
Excessive Collateral Branching of Axons at the Lesion Site

Injury to the peripheral nerve sets initiates a complex series of changes distal to the site of injury, collectively known as Wallerian degeneration. Within 24 h after lesion, the axonal content begins to necrotize and axonal debris is phagocytosed by blood-borne macrophages and proliferated Schwann cells (Perry and Brown 1992; Hirata and Kawabuchi 2002; McPhail et al. 2004a). When resorption is complete, the Schwann cells form long chains of cells (bands of Büngner), which bridge the interfragmentary gap and form guiding channels for the regenerating branches on their way to the target(s). The architectural pattern of the Büngner's bands of the peripheral stump remains unchanged for 3 months, after which progressive distorsion by proliferating connective tissue occurs. The process of Wallerian degeneration creates an environment that is highly supportive for axonal growth. The preference for axonal growth into a degenerating nerve ensures that the vast majority of axons will regrow into the distal stump, if it remains in continuity with the proximal stump (Bisby 1995).

In spite of that, the regenerating axons do not merely elongate toward the distal stump, but respond with axonal branching (sprouting) by lateral budding mainly at the nodes of Ranvier, up to 6 mm proximal to the injury site. As regeneration proceeds, some of these supernumerary branches are pruned off over a period of up to 12 months (Bray and Aguayo 1974). There are, however, persistently higher numbers of myelinated and unmyelinated axons in regenerated segments of peripheral nerves than in intact nerves. Axonal branching begins from the end-bulb within 3 h after injury (Sjoberg and Kanje 1990). The regenerating branches initially lie on the surface of the Schwann cells. Later, these branches increase in diameter and get surrounded by Schwann cell processes.

Observations in vitro show that axonal branching begins from the end-bulb within 3 h after injury (Sjoberg and Kanje 1990). The regenerating branches initially lie on the surface of the Schwann cells. Later, these branches increase in diameter and get surrounded by Schwann cell processes. The guidance of these immature axons to their final destination can be considered as a series of short-range projections to intermediate targets under the influence of local guidance cues (see below). Neurons respond to these cues by means of motile sensory apparatus at the tip of the advancing axon termed the *"growth cone,"* which very often does not emerge from the axon at the precise site of injury, but proximal to it (Ziv and Spira 1997). The initial formation of growth cones occurs before the necessary newly synthesized proteins would have time to arrive at the site of axon injury, that is, too rapidly to be dependent on metabolic changes in the cell body (Smith and Skene 1997).

The growth cone borne by neurites is shaped like a webbed foot (Fawcett and Keynes 1990). There is a swollen central core from which flattened processes called *lamellipodia* and numerous stiff fine processes called *filopodia* extend. Current studies have identified three major intracellular cytoskeletal components responsible for the cytomechanical forces in the leading edge of elongating axons: actin microfilaments, myosin, and microtubules (Challacombe et al. 1996). The growth cone formation begins with a restructuring of the neurofilaments and microtubules to form an altered cytoskeletal region proximal to the tip of the transected axon in which vesicles accumulate. This rearrangement of the cytoskeleton forms a transient cellular compartment that traps the transported vesicles and serves as a locus for microtubule polymerization. Microtubuli, in turn, facilitate the fusion of vesicles with the plasma membrane, promoting the extension of growth cone lamellipodia (Spira et al. 2003).

The recognition of specific guiding cues is performed by the actin-rich filopodia, which have a guidance and/or sensory role, sniffing out gradients of trophic or adhesive factors (Lin and Forscher 1993). Isolated filopodia can respond to alterations in their environment by changes in internal calcium concentrations, and filopodia on different parts of the growth cone respond independently (Bixby and Harris 1991; Letourneau and Cypher 1991; Gordon-Weeks 1997).

1.3
Role of Cytoskeleton Reorganization During Axonal Regrowth

In response to axotomy, the synthesis of cytoskeletal proteins in the perikarya is increased (Hoffman and Lasek 1980). A postaxotomy increase in overall tubulin synthesis has been documented (Oblinger and Lasek 1988), and it is thought that upregulated levels of tubulin in the perikarya and increased delivery of mictotubules to regrowing axon tips are essential for effective regeneration after injury (Tetzlaff et al. 1988a, 1991, 1996).

The rate of elongation of an axon is determined by the rate at which the growth cone can advance over the substrate. In rat sciatic nerve, both large and small diameter sensory axons elongate at nearly the same rate as do somatic motor axons (about 4 mm/day; Fawcett and Keynes 1990). In the regenerating (crushed) facial nerve of rats, the rate of axonal elongation is 4.3 mm/day measured from the transport of radiolabeled protein (Tetzlaff and Bisby 1989).

Axonal elongation depends on the advance of microtubules that provide structural support and serve as tracks for axonal transport of membraneous organelles. Stable microtubule bundles project from the axon into the central region (C-domain) of the growth cone, whereas the ends of dynamic microtubules expand and stretch into the actin-rich P-domain (Gordon-Weeks 1991). Goldberg and Burmeister (1986) and Aletta and Greene (1988) have described three phases of axonal elongation. First, lamellipodia and filopodia are extended from the tip of the axon (*protrusion*). Second, microtubules enter the recently protruded regions of the growth cone (*engorgement*). Third, the portions of the growth cones lateral to the engorged regions become quiescent and coalesce to form a new portion of the axon (*consolidation*).

The net protrusion of lamellae and filopodia is largely determined by the rates of F-actin polymerization and retrograde flow (Lin et al. 1994). If actin polymerization is blocked, leading edge protrusion does not occur and F-actin is removed from the peripheral (P) domain by retrograde transport. On the other hand, if F-actin retrograde flow is inhibited, then the rate of protrusion of the leading edge will be determined primarily by the polymerization of F-actin. Rho-family GTPases (Rho, Rac, Cdc42) have been found to mediate the formation of filopodia and lamellipodia, that is, to be involved in axon guidance (see Gallo and Letourneau 1998 for review) and also in growth cone responses to collapsing guidance cues (Jin and Strittmatter 1997).

Results from some additional experiments have suggested that axonal growth requires microtubules (both addition of tubulin to polymer and transport of preestablished polymer) at the growth cone (Yu and Baas 1995; Baas 1997, 1999). Tanaka and Kirschner (1991, 1995) report that microtubules in growth cones appear to be transported by "pushing" toward the leading edge of the P-domain. Consistent with this interpretation, Challacombe et al. (1997) report that looped microtubules in growth cones stain with a marker for stable microtubule polymer (i.e., detyrosinated α-tubulin). Therefore, both microtubule polymerization and transport contribute to axonal elongation by advancing microtubules into the P-domain of the growth cones.

Still, the exact nature of F-actin–microtubule interactions in the axon growth cone is not well understood. Growth cones at the tips of rapidly extending axons are small and highly active. However, in preparation for branching, they may pause for many hours, greatly enlarge, and maintain motility without a forward advance. Subsequently, a new growth cone develops from the tip of the large pausing growth cone and forms a new leading axon. Remnants of the large pausing growth cone remain on the axon shaft as filopodial and lamellar expansions that subsequently give rise to axon collaterals (Halloran and Kalil 1994; Szebenyi et al. 1998).

Microtubules in the central region of advancing growth cones get stretched out. In slowly growing axons, microtubules become bundled and in pausing growth cones - they form prominent loops (Tanaka and Kirschner 1991). Transition to new axonal growth and branch formation is accompanied by splaying of looped microtubules and formation of short microtubule fragments that invade the lamellipodium (Dent et al. 1999). Thus, growth cone pausing is closely related to the mechanism of branching (Dent and Kalil 2001).

Within the axon, the microtubule array is continuous from the cell body into the terminal growth cone, but individual microtubules vary in length, stopping and starting at various points within the array (Bray and Bunge 1981; Yu and Baas 1994). All microtubules have a consistent 13-protofilament lattice (Tilney et al. 1973; Burton et al. 1975) and are uniformly oriented with regard to their intrinsic polarity, with plus end directed away from the cell body (Heidemann et al. 1981; Baas et al. 1988).

Axons branch principally by the formation of collaterals rather than by bifurcation of the terminal growth cone (O'Leary and Terashima 1988). The generation of axon collateral branches involves a reinitiation of cell surface motility from regions of the axons that have been quiescent (Bastmeyer and O'Leary 1996). The first step of axon collateral branch formation involves the protrusion of filopodia from the axon shaft (Yu et al. 1994). Most of these filopodia have a short lifetime, but a subset becomes stabilized by the entry of stable, though few, microtubules and continues to grow developing into collateral branches that can reach a significant length (Dent et al. 1999).

Results of Yu et al. (1994) show that the region of the parent axon, from which the collateral branch forms, contains about 20% less polymer as compared to regions of parent axon not forming a branch. Moreover, there are 10 times as many free microtubule ends and the microtubules on average are about 10 times shorter. The microtubules within the newly formed collateral branches are on average the same as within the parent axon, indicating that these microtubules were assembled in the parent axon and then transported into the branch. These observations provide strong support for the view that there is a local fragmentation of the microtubules during collateral branch formation.

1.4
Exchange of Nerve Impulses Between Adjacent Axons

After injury, each parent axon may give rise to 25 daughter axons (Shawe 1954; Jenq et al. 1988). As regeneration proceeds, some of these supernumerary branches are pruned off over a period of up to 12 months (Mackinnon et al. 1991; Brushart et al. 1998). Those that are lost are presumably those that fail to make a connection with a peripheral target. There are, however, persistently higher numbers of myelinated and unmyelinated axons in regenerated segments of peripheral nerves than in the corresponding parent nerves (Horch and Lisney 1981; Murphy et al. 1990).

The excessive firing by the transected axons is a consequence of trans-axonal exchange of abnormally intensive nerve impulses (ephaptic cross talk) between axons from adjacent fascicles (Sadjadpour 1975). This usually occurs when axonal forward growth is blocked and the branches are stunted forming a tangled terminal mass (a "neuroma"). The growth process and the steering of the cones is further complicated by the presence of branches from the distal nerve stump (Shaw and Bray 1977) and by collateral branches of nearby intact nerve fibers (Diamond et al. 1987). The initially formed growth cones transform into swollen "end-bulbs" and form disseminated "microneuromas" scattered along the distal nerve trunk, its branches, and its target tissue. After about 1 week, these neuromas begin to discharge action potentials spontaneously, perhaps as the result of the concentration of large numbers of sodium channels (Devor et al. 1989).

1.5 Vigorous Terminal Sprouting of Axons in the Denervated Muscles

Another problem is the intramuscular sprouting of axons. Upon reaching a target, axons undergo additional sprouting to reinnervate many muscle fibers and thus form new and larger motor units (Grimby et al. 1989; Trojan et al. 1991; Son et al. 1996; Tam and Gordon 2003; Gordon et al. 2004). Sprouting, thus, has also a "maladaptive" side leading to reinnervation of motor endplates by more than one motoneuron, a state known as "polyinnervation" (Brown et al. 1981; Rich and Lichtman 1989). The performance of a muscle fiber controlled by two or more asynchronously firing motoneurons is not physiologically advantageous.

Guntinas-Lichius et al. (2005) concentrated on the intramuscular axonal sprouting and compared the reinnervation pattern of m. levator labii superioris in visually normal rats (no recovery of whisking after injury) to that in blind animals (with complete restoration of vibrissae motor performance; Tomov et al. 2002). They found that motor endplates with morphological signs of multiple innervation were much less frequent in the blind rats (10% of all endplates) as compared to the sighted animals (51%). As an accompanying event to the probably reduced intramuscular sprouting in the blind animals, Peeva et al. (2006) described a very strong expression of neuronal class III β-tubulin in the regrowing axons.

1.6 Cellular Correlates of Muscle Reinnervation: the Role of Terminal Schwann Cells

Denervated Terminal Schwann Cells (TSC) can enlarge and sprout processes, which reach adjacent innervated motor endplates (Reynolds and Woolf 1992; O'Malley et al. 1999; Griffin and Thompson 2008). Using these bridges, TSC reach, attract, and direct intramuscular axonal sprouts toward the denervated endplates (Love and Thompson 1999; Kang et al. 2003; Reddy et al. 2003). Interestingly enough, it has been shown that the outgrowth of TSC processes precedes the outgrowth of sprouts from the intact intramuscular axons, that is, TSC are able to initiate intramuscular axonal sprouting (Son and Thompson 1995). Thus, the beneficial effect of stimulation on muscle reinnervation may be mediated by interfering with the extension of TSC processes and their ability to bridge between the endplates. Similar results about perturbed formation of TSC bridges, though after running exercise (Tam and Gordon 2003) or electrical stimulation (Love et al. 2003), have been recently reported. Thus, any form of artificially excited muscular activity may inhibit the bridge formation by TSC and reduce postlesional intramuscular sprouting.

1.7 Molecular Correlates of Muscle Reinnervation: Role of Sprouting-Inducing Stimuli

Reduced amounts of sprouting-inducing stimuli could explain the low portion of polyinnervated motor endplates detected in the vibrissal muscles (e.g., m. levator labii superioris) after mechanical stimulation (Guntinas-Lichius et al. 2005). Denervated muscles have been shown to produce short-range diffusible sprouting stimuli (Slack and Pockett 1981; Pockett and Slack 1982; English 2003; Zhao et al. 2004). Various neurotrophic factors have been identified as possible candidates for this role (Sendtner 1998; Raivich and Makwana 2007). Their amount is inversely proportional to muscle activity (Brown and Ironton 1977; Brown et al. 1981).

1.8 Questions Still Open

Peripheral nerve regeneration is a pointless process unless the regenerating axons grow back to reinnervate their original muscle targets. The inevitable "postparalytic syndrome," including mass movements (synkinesia) and altered reflexes (Baker et al. 1994), has been attributed to the "misdirected" reinnervation of the targets (Sumner 1990). The misdirected reinnervation can be compared to a chain of processes, which take place at two locations.

1. At the lesion site, axons may be misrouted along the "wrong" nerve fascicle due to an insufficient and/or malfunctioning axonal guidance. In consequence, a denervated muscle receives reinnervation by "alien" axons (Aldskogius and Thomander 1986).
2. At the neuromuscular junction (NMJ), a muscle fiber can be reinnervated by several motoneuronal axons or axonal branches (polyinnervation;

Gorio et al. 1983; Fu and Gordon 1997) most probably due to the presence of competing supernumerary branches from all transected axons (Dyck and Hopkins 1972).

In an attempt to shed more light on several key processes in this chain, we tried to provide reasonable answers to the following questions:

1. Can experimentally reduced collateral branching of axons at the lesion site, for example, by means of local neutralization of trophic factors, be associated with improved recovery of function?
2. Is there a relationship between the amount of neurite growth-related proteins (actin, tubulin) and improved recovery of function?
3. If yes, is the increased amount due to an increased synthesis, due to a slower transport, or due to delayed turnover of tubulin?
4. Would therapy with brief, low-frequency intraoperative electrical stimulation (IOES; 1 h, 20 Hz) promote successful recovery of function after transection and suture of the facial nerve (facial–facial anastomosis, FFA)?
5. Would therapy with postoperative electrical stimulation (POES) of the vibrissal muscles promote successful recovery of function after FFA?
6. Would therapy with postoperative manual mechanical stimulation of the vibrissal muscles promote successful recovery of whisking function after FFA?
7. Would therapy with postoperative manual mechanical stimulation of the vibrissal muscles promote successful recovery of whisking function after transection of the facial and hypoglossal nerves and suture of the proximal stump of the hypoglossal nerve to the distal fragment of the facial nerve (hypoglossal–facial anastomosis, HFA)?
8. Would therapy with postoperative manual mechanical stimulation of the vibrissal muscles promote successful recovery of whisking function after interpositional facial nerve grafting (IPNG)?
9. Would therapy with postoperative manual mechanical stimulation of the orbicularis oculi muscle (OOM) promote successful recovery of eye-closure after FFA?
10. Would therapy with postoperative manual mechanical stimulation of the tongue muscles promote successful recovery of function after transection and suture of the hypoglossal nerve (hypoglossal–hypoglossal anastomosis, HHA)?
11. Would therapy with postoperative manual mechanical stimulation of the forearm muscles promote successful recovery of grasping function after transection and suture of the median nerve (medianus–medianus anastomosis, MMA)?
12. Do trigeminal sensory afferents play a role in the accomplishment of the effects of manual mechanical stimulation?

1.9 Methodological Approach

The answers to these questions have been sought by an extensive and combined methodological approach consisting of:

1. *Video-based motion analysis of whisking behavior* which provides a very sensitive tool to study the motor recovery of muscles innervated by the facial nerve (Tomov et al. 2002).
2. *Simultaneous multiple fluorescent neuronal labeling to quantitatively estimate the degree of axonal branching* (Angelov et al. 1999; Dohm et al. 2000; Streppel et al. 2002).
3. *Combined staining of axons* (antineuronal class III β-tubulin) *and neuromuscular junctions* (AlexaFluor 488-conjugated α-bungarotoxin) to estimate the quality of target muscle reinnervation (Guntinas-Lichius et al. 2005; Angelov et al. 2007).

Chapter 2
Attempts to Improve Axonal Pathfinding and Quality of Target Reinnervation

Numerous experiments were grouped into three major sets.

In the first major set, we report our attempts to improve axonal pathfinding by reduction of collateral axonal branching at the lesion site by means of (1) local trophic factor neutralization or (2) application of established pharmacological agents to perturb microtubule assembly.

In the second major set, we describe several experiments in which we tried to reduce intramuscular axonal sprouting in denervated muscles by (1) modification of microtubule dynamics, (2) intraoperative electrical stimulation (IOES) of the transected facial nerve before its surgical reconstruction, (3) postoperative electrical stimulation (POES) of denervated vibrissal muscles, (4) manual mechanical stimulation of denervated vibrissal muscles after varying facial nerve reconstruction strategies, (5) manual mechanical stimulation of denervated orbicularis oculi muscle (OOM), (6) manual mechanical stimulation of denervated tongue muscles, and (7) manual mechanical stimulation of denervated forearm muscles.

In the third major set, we summarize the results from experiments in which we tried to reveal the mechanism of action of the manual mechanical stimulation by means of experiments with permanent abolishment of the trigeminal afferents from the whisker pad. Before and after the experiments, all animals were kept on standard laboratory food (Altromin, 32791 Lage, Germany) and tap water ad libitum with an artificial light–dark cycle of 12 h light on, 12 h off. We used female young adult (175–200 g) Wistar rats (strain HsdCpb:WU, Harlan-Winkelmann, Borchen, Germany), because testosterone has been shown to beneficially affect peripheral nerve regeneration (Yu and Yu 1983).

All experiments were conducted in accordance with the "German Law for Animals Protection" and were approved by the local animal care committee (Bezirksregierung Köln).

For a given parameter, data from all relevant experimental groups were tested in one-way analysis of variance (one-way ANOVA) procedure for overall experimental effects. If significant effects were detected ($p < 0.05$), comparisons of all groups with one control group were performed using the post hoc test of Dunnett at a significance level of 0.05. For analysis, Statistica 6.0 software (StatSoft, Tulsa, OK, USA) was used.

2.1 Efforts to Reduce Collateral Axonal Branching at the Lesion Site

2.1.1 Neutralization of Trophic Factors at the Lesion Site Reduced Collateral Axonal Branching, but Did Not Improve Recovery of Function

Within an earlier experimental set, Streppel et al. (2002) tested the hypothesis that neutralization of diffusable neurotrophic factors (NGF, BDNF, FGF-2, IGF-1, CNTF, GDNF) at the lesion site could reduce collateral branching of transected axons and thus improve quality of reinnervation. Neutralizing concentrations of anti-NGF, anti-BDNF, and anti-IGF-I significantly reduced branching. These results appeared very promising with regard to the feasibility of the method for treatment of patients, on one side, and to the presumed dominant contribution of axonal misguidance to the failure of recovery, on the other. However, experimental data proving that suppressed branching would enhance precision of target reinnervation and favor functional recovery were required. In other words, before considering this treatment as of potential value in clinical situations it should be known whether reduced collateral axonal branching at the lesion site would be associated with improved recovery of motor function.

The results from all counts in this chapter were obtained in a manner that was absolutely identical (rat strain, sex, weight, surgical treatment, postoperative survival period, tracers, etc.) to that used in an earlier set of experiments (Streppel et al. 2002; Angelov et al. 2005). This permits pooling of the present and earlier data. The larger sample size in each group (17 instead of eight animals) allowed a more powerful statistical analysis and more reliable conclusions.

2.1.1.1 Animals

Seventy-two Wistar rats were distributed in nine groups (group 1 of intact controls and groups 2–8 of surgically treated) each consisting of eight animals.

In addition, 12 female young adult Sprague-Dawley (SD) rats were divided into two groups. Half of the animals were with normal visual perception (purchased from Charles River, Germany) and the other group consisted of blind rats (substrain Royal College of Surgeons, RCS, generously supplied by U. Schraermeyer). The SD/RCS rats lose their photoreceptor cells 2 months after birth due to a genetic defect of the retinal pigment epithelium (D'Cruz et al. 2000; Sheedlo et al. 1991). Thus, these animals can obtain spatial information only by their mystacial vibrissae (Brecht et al. 1997).

2.1.1.2 Overview of Experiments

Two months after surgery on the facial nerve, the recovery of vibrissal motor performance was assessed in all surgically treated rats (84 animals) using video-based motion analysis (VBMA) of explorative whisking. Thereafter, the animals of groups 1–9 were used to analyze the degree of posttransectional collateral axonal branching by retrograde labeling with three different fluorescent dyes applied simultaneously to three different branches of the facial nerve. All Sprague-Dawley rats (group SD and group SD/RCS) were not subjected to tracer applications, but used to study qualitative and quantitative aspects of muscle reinnervation by means of immunocytochemistry for neuronal class III β-tubulin and histochemistry with alpha bungarotoxin.

2.1.1.3 Facial–Facial Anastomosis

Transection and end-to-end suture of the right facial nerve (facial–facial anastomosis, FFA) were performed in group 2 of the Wistar rats and in all Sprague-Dawley animals (group SD and group SD/RCS). Following an intraperitoneal injection of ketamin/xylazin, the main trunk of the facial nerve was exposed and transected close to its emergence out of the foramen stylomastoideum (Fig. 2.1a). The proximal stump was then microsurgically reconnected to the distal stump with two 11–0 atraumatic sutures (Ethicon, Norderstedt, Germany). Finally the wound was closed by three 4–0 skin sutures (Ethicon).

2.1.1.4 Entubulation of the Facial Nerve Trunk

Under anesthesia, the right facial nerve in groups 3–9 of the Wistar rats was transected and both stumps were inserted into a silicone precision tube (Fig. 2.1b). The tube had an inner diameter of 1.47 mm and outer diameter of 1.96 mm (Aromando Medizintechnik, Düsseldorf, Germany). The space between the proximal and distal nerve stumps with a volume of approximately 8 μl (5 mm $\times$ 0.735 mm $\times$ 0.735 mm $\times$ π) was filled with collagen type I (group 3) or with collagen gel containing antibodies to trophic factors in the following neutralizing concentrations (Fig. 2.1b):

Group 4: mouse monoclonal anti-NGF (40 μg/ml; Roche, Mannheim, Germany; Bedi et al. 1992; Diamond et al. 1992; Ro et al. 1996)

Group 5: mouse monoclonal anti-BDNF (160 μg/ml; R&D Systems, Wiesbaden, Germany; Tonra et al. 1998)

Group 6: mouse monoclonal anti-bFGF (100 μg/ml; UBI/Biomol, Hamburg, Germany; Tuttle et al. 1994; Murai et al. 1996)

Group 7: mouse monoclonal anti-IGF-I (30 μg/ml; UBI/Biomol, Zheng et al. 1997)

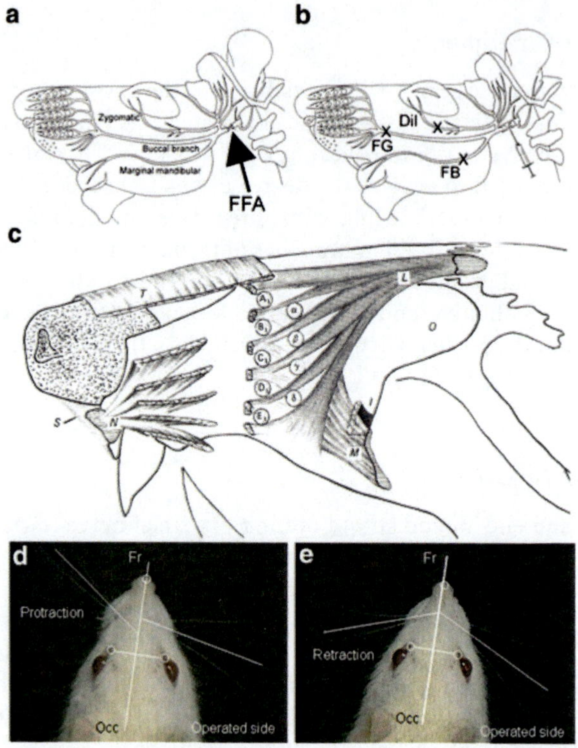

Fig. 2.1 (**a–e**) Schematic drawings of the infratemporal portion of the rat facial nerve. The site of transection and end-to-end suture of the facial nerve trunk, i.e., facial–facial anastomosis (FFA) is indicated by an *arrow* (**a**). The site of entubulation and the tracer application are indicated by abbreviations of the three different labels applied, i.e., DiI, FG, and FB, respectively (**b**). Schematic drawing of the extrinsic vibrissae muscles according to Dörfl (1982): α-δ: the four caudal hair follicles, the muscles slings of which "straddle" the five vibrissae rows (**a–e**); T – m. transversus nasi; L – m. levator labii superioris; N – m. nasalis; M – m. maxillolabialis; O – orbit; S – septum intermusculare (**c**). Analysis of the vibrissae motor performance with precise measurement of angles, angular velocity, and angular acceleration of the intact and operated side during protraction (**d**) and retraction (**e**) of the vibrissae. Note the significant change in angle between the sagittal line **Fr-Occ** during protraction and retraction on the intact side. The vibrissae on the operated side remain stiff. Adopted from Guntinas-Lichius et al. 2002

Group 8: mouse monoclonal anti-GDNF (3 μg/ml; R&D Systems, Wiesbaden; Vega et al. 1996)

Group 9: goat polyclonal anti-CNTF (100 μg/ml; R&D Systems; Ding et al. 1994; Tokiwa et al. 1994)

Since, except for anti-CNTF, all neutralizing antibodies were raised in mice, an entubulation of the facial nerve in a gel containing mouse nonimmune IgG would be a necessary control. However, previous own experiments have shown that the application of mouse nonimmune IgG (160 µg/ml; Sigma) has no reducing effect on axonal sprouting (Streppel et al. 2002). Therefore, such a control group was not included in this study.

2.1.1.5

Observations on the Function of Mystacial Vibrissae

The ability to observe postoperative paralysis of vibrissae and the subsequent gradual recovery of rhythmical whisking is one of the major advantages of the "facial-nerve transection model." Under normal physiological conditions, the mystacial vibrissae of the rat are erect with anterior orientation. Their simultaneous sweeps known as "whisking" or "sniffing" (Semba et al. 1980; Welker 1964) occur 5–11 times per second (Bermejo et al. 1996; Komisaruk 1970; Carvell and Simons 1990). The key movements of this motor activity are the protraction and retraction of the vibrissal hairs by the piloerector (follicular) muscles. The striated muscle fibers mediating protraction form a sling around the rostral aspect of each hair follicle: contraction of these muscles pulls the base of the follicle caudally, moving the distal aspects of the whisker hair forward. By contrast, retraction of the vibrissae depends primarily upon passive elastic properties of the deep connective tissue (Dörfl 1985; Wineski 1985). All muscles are innervated by the buccal and marginal mandibular branches of the facial nerve (Dörfl 1985).

As a result of muscle denervation after nerve transection, the whiskers acquired a caudal orientation and remained motionless within the first 10 days after surgery. At 10–14 days postoperation (DPO), the vibrissae rose to the level of the mouth with a posterior orientation in all animals irrespective of treatment. An overall poor restoration of rhythmical whisking was observed in all experimental groups.

2.1.1.6

Analysis of Vibrissae Motor Performance

The method of VBMA of vibrissae motor performance has been thoroughly established and tested in a series of experiments (Tomov et al. 2002). Two months after surgery on the facial nerve, all animals (of the strains Wistar, SD and SD/RCS) were videotaped for 3–5 min during active exploration using a digital camcorder (Panasonic NV DX-110 EG). Selected sequences containing the most pronounced movements of the intact contralateral vibrissal hairs were captured by a 2D/Manual Advanced Video System (PEAK Motus 2000, PEAK Performance Technologies, Inc., Englewood, CO, USA). The geometrical model consisted of three reference points (1) a point in the medial sagittal line close to the end of the nose, (2) a point corresponding to the medial angle of the left orbita, and (3) a point corresponding to the medial angle of the right orbita (Fig. 2.1d, e). Using this model, we collected and evaluated data on the following parameters:

- Whisking frequency: cycles of protraction (forward movement) and retraction (backward movement) per second
- Angle at maximal protraction: the rostrally open angle between the mid-sagittal plane and the hair shaft (in degrees)
- Amplitude: the difference between maximal retraction and maximal protraction (in degrees)
- Angular velocity during protraction (in degrees per second)
- Angular acceleration during protraction (in degrees per second^2)

2.1.1.7
Intact Rats

During exploration the mystacial vibrissae swept back and forth with a frequency of about 6 Hz. The maximal protraction (a rostrally open angle between the vibrissa shaft and the median sagittal plane) was about 70°. The mean amplitude of whisking (the difference between maximal retraction and maximal protraction in degrees) measured about 50°. These movements were performed at a sagittal angular velocity of about 500°/s and a sagittal angular acceleration of 20,000°/s^2 (Table 2.1, group intact). These results are consistent with our previous observations (Tomov et al. 2002).

2.1.1.8
Operated Rats

Also consistent with previous data are the results obtained from animals with facial nerve anastomosis (Table 2.1, group 2; Tomov et al. 2002). As compared with intact animals, large functional deficiency was evident from the significantly larger angle at maximal protraction (+28% vs. the *Intact* group), the smaller amplitude of vibrissae movement (−72%), as well as the lower angular velocity and acceleration during protraction (−75% and 87%, respectively). These post-operative changes were due to inadequate muscle function during the active protraction phase (see above). The frequency of movements was similar to intact animals which is explainable by the roughly similar reduction of both range and speed of movement, as well as by the influence of passive elastic tissue properties on this parameter.

Taken together, these findings show that the range and velocity of movements remained severely impaired even 2 months after facial anastomosis. As compared with the FFA group, application to the transected and entubulated facial nerve of collagen alone or collagen plus neutralizing antibodies did not produce any

Table 2.1 Motor recovery after entubulation of the facial nerve into silicone tubes containing neutralizing antibodies against neurotrophic factors

Group of animals	Frequency (in Hz)	Angle at maximal protraction (in degrees)	Amplitude (in degrees)	Angular velocity during protraction (in degrees/s)	Angular acceleration during protraction (in degrees/s^2)
1. Intact	6.4 ± 1.1	71 ± 15*	50 ± 15*	530 ± 330*	19,293 ± 13,514*
2. FFA	6.3 ± 0.5	91 ± 12	19 ± 6	135 ± 54	2,485 ± 792
3. Collagen	5.0 ± 0.9	91 ± 12	15 ± 5	146 ± 37	3,036 ± 1,576
4. NGF	5.3 ± 0.9	94 ± 13	16 ± 5	113 ± 64	2,648 ± 2,202
5. BDNF	5.9 ± 0.5	93 ± 12	17 ± 6	152 ± 77	3,176 ± 2,109
6. bFGF	5.8 ± 1.3	103 ± 16	18 ± 9	193 ± 50	3,417 ± 1,643
7. IGF-I	5.8 ± 1.7	93 ± 18	17 ± 8	283 ± 140	2,322 ± 1,381
8. GDNF	5.4 ± 1.4	99 ± 22	16 ± 6	245 ± 122	2,600 ± 1,350
9. CNTF	4.6 ± 0.7*	113 ± 6*	8 ± 2*	164 ± 20	816 ± 294

Biometrics of vibrissae motor performance in intact rats (group *Intact*), in rats after transection and suture of the right facial nerve (*FFA*), and after entubulation of the facial nerve into a silicone tube containing collagen Type I alone (collagen) or collagen plus antibodies against neurotrophic factors (NGF, BDNF, bFGF, IGF-I, GDNF, or CNTF) in neutralizing concentrations (see Material and Methods). Each group consisted of eight animals (Wistar rats, strain HsdCpb:WU). Shown are group mean values ± SD. Group mean values significantly different from the control group (2. FFA, ANOVA and post hoc Dunnett's test, $p < 0.05$) are indicated by asterisks

positive functional effects (Table 2.1, groups *Collagen, NGF, BDNF, IGF-I and GDNF*). Since treatment with some antibodies, however, significantly reduced axonal branching (Table 2.2), we suggested that the excessive collaterals at the lesion site may not be a critical factor in the recovery of coordinated muscle activity. As an attempt to search for this critical factor, we decided to estimate the quality of target muscle reinnervation.

2.1.1.9

Simultaneous Application of Three Fluorescent Tracers (DiI, FG, and FB)

One day after videotaping, the animals received an intraperitoneal injection of ketamin/xylazin. The zygomatic, buccal, and marginal mandibular ramus of eight intact (group 1) and 64 operated (groups 2–9) animals were transected and instilled with crystals of DiI (1,1′-dioctadecyl-3,3,3′,3′-tetramethylindocarbocyanine perchlorate, Molecular Probes, Leiden, The Netherlands), Fluoro-Gold (FG; Fluorochrome Inc., Denver, CO, USA), and Fast Blue (FB; EMS-Chemie GmbH, Groß-Umstadt, Germany), respectively (Fig. 2.1b). To avoid the mingling of FG and FB (both water soluble) and DiI, all crystals were left in situ only for 30 min. Thereafter, the application site was carefully rinsed and dried and the wound was closed.

Table 2.2 Degree of collateral axonal branching after treatment of the transected facial nerve with neutralizing antibodies against neurotrophic factors

Group of animals	Neurons projecting only into the zygomatic nerve (DiI-only)	Neurons projecting into the zygomatic and buccal nerves (DiI + FG)	Neurons projecting into zygomatic and marginal mandibular nerves (DiI + FB)	All DiI labeled neurons projecting into the zygomatic nerve (DiI, DiI + FG, DiI + FB)	Neurons projecting only into the buccal nerve (FG-only)	Neurons projecting only into the marginal mandibular nerve (FB-only)
1. Intact	364 ± 47* 100%	–* 0%	–* 0%	364 ± 47* 100%	1,441 ± 101*	379 ± 94*
2. FFA	213 ± 53 30%	239 ± 52 34%	257 ± 56 36%	709 ± 178 100%	1,908 ± 289	1,488 ± 356
3. Collagen	194 ± 89 26%	334 ± 74* 44%	227 ± 74 30%	755 ± 125 100%	1,966 ± 203	1,543 ± 348
4. NGF	343 ± 78* 70%	74 ± 32* 16%	63 ± 59* 14%	465 ± 133* 100%	1,408 ± 562*	546 ± 156*
5. BDNF	360 ± 73* 74%	79 ± 71* 16%	49 ± 40* 10%	488 ± 125* 100%	1,538 ± 610	565 ± 204*
6. bFGF	321 ± 95* 79%	50 ± 47* 13%	30 ± 28* 8%	401 ± 76* 100%	1,200 ± 672*	580 ± 148*
7. IGF-I	361 ± 81* 64%	118 ± 92* 20%	98 ± 42* 16%	578 ± 140* 100%	1,635 ± 616	652 ± 143*
8. GDNF	361 ± 98* 67%	131 ± 66* 24%	49 ± 30* 9%	541 ± 127* 100%	1,633 ± 575	668 ± 135*
9. CNTF	348 ± 69* 71%	109 ± 22* 23%	29 ± 10* 6%	482 ± 70* 100%	1,425 ± 312	701 ± 141*

Number of retrogradely labeled motoneurons with projection axons in the zygomatic, buccal, or marginal mandibular branches of the facial nerve of intact rats (group Intact), after transection and suture of the facial nerve (FFA), and after entubulation with collagen Type I (collagen) or collagen plus antibodies against NGF, BDNF, bFGF, IGF-I, or GDNF. Seventeen animals (Wistar rats, strain HsdCpb:WU) were studied per group. Shown are group mean values ± SD. Group mean values significantly different from the control group (FFA, ANOVA and post hoc Dunnett's test, $p < 0.05$) are indicated by asterisks. The percentage values below the absolute numbers in columns 2–5 indicate the proportions of motoneurons projecting through the zygomatic nerve with branched axons (DiI + FG or DiI + FB, column 3 and 4) and unbranched axons (DiI-only, column 2) are indicated as percentages of all neurons carrying DiI label (column 5)

2.1.1.10

Tissue Preparation

Ten days after triple labeling, all animals were fixed by perfusion with 4% paraformaldehyde in 0.1 M phosphate buffer pH 7.4. Their brainstems were cut in 50 μm thick vibratome serial sections.

2.1.1.11
Fluorescence Microscopy After Triple Retrograde Labeling

Sections were observed with a Zeiss Axioskop 50 epifluorescence microscope through a custom-made band pass-filter set (no. F31-000; excitation D 436/10; beam splitter 450 DCLP; barrier filter D470/40), which allows recognition only of FB-labeled (appearing blue). Observations through a custom-made HQ-Schmalband-filter set (no. F36-050; excitation D 369/40; beam splitter 400DCLP; barrier filter HQ 635/30) and a HQ-Schmalband-filter set for Fluoro-Gold visualized all motoneurons containing FG. Observations through a filter set 15 of Carl Zeiss (Excitation BP 546/12, Emission LP 590) revealed the red fluorescence of those motoneurons retrogradely labeled by DiI. The fluorescence crosstalk between the tracers was restricted with this filter combination *ad maximum*:

2.1.1.12
Image Analysis After Triple Retrograde Labeling

Employing a CCD Video Camera System combined with the image analyzing software Optimas 6.5. (Optimas Corporation, Bothell, Washington, DC, USA), separate color images of retrogradely labeled facial motoneurons were created through the different filter sets, and all cells stained by DiI_{only}, FG_{only}, FB_{only} and all cells double stained by DiI + FG or DiI + FB were identified and manually counted on the computer screen as described by Dohm et al. (2000).

2.1.1.13
Qualitative and Quantitative Estimates in the Facial Nucleus of Intact Rats

Motoneurons innervating muscles through the zygomatic, buccal, or marginal mandibular branches are localized in three distinct subnuclei of the facial nucleus – the dorsal, lateral, and intermediate subnucleus, respectively (Hinrichsen and Watson 1984; Thomander 1984; Klein and Rhoades 1985; Aldskogius and Thomander 1986; Semba and Egger 1986; Ito and Kudo 1994; Dohm et al. 2000; Fig. 2.2a). The average size of these distinct populations, as estimated by the triple labeling in this experiment, was around 360 zygomatic, 1,440 buccal, and 380 marginal mandibular motoneurons (Table 2.2, group *Intact*). These data are consistent with previous reports (Dohm et al. 2000; Streppel et al. 2002). Double labeling of zygomatic, buccal, and marginal mandibular motoneurons was not observed (Fig. 2.2a; Table 2.2), which is consistent with the myotopic principle and indicates the consistent reproducibility of the tracing technique. The latter notion is further supported by the observation that no labeled cells were found in the medial or ventromedial facial subnuclei which contain motoneurons projecting

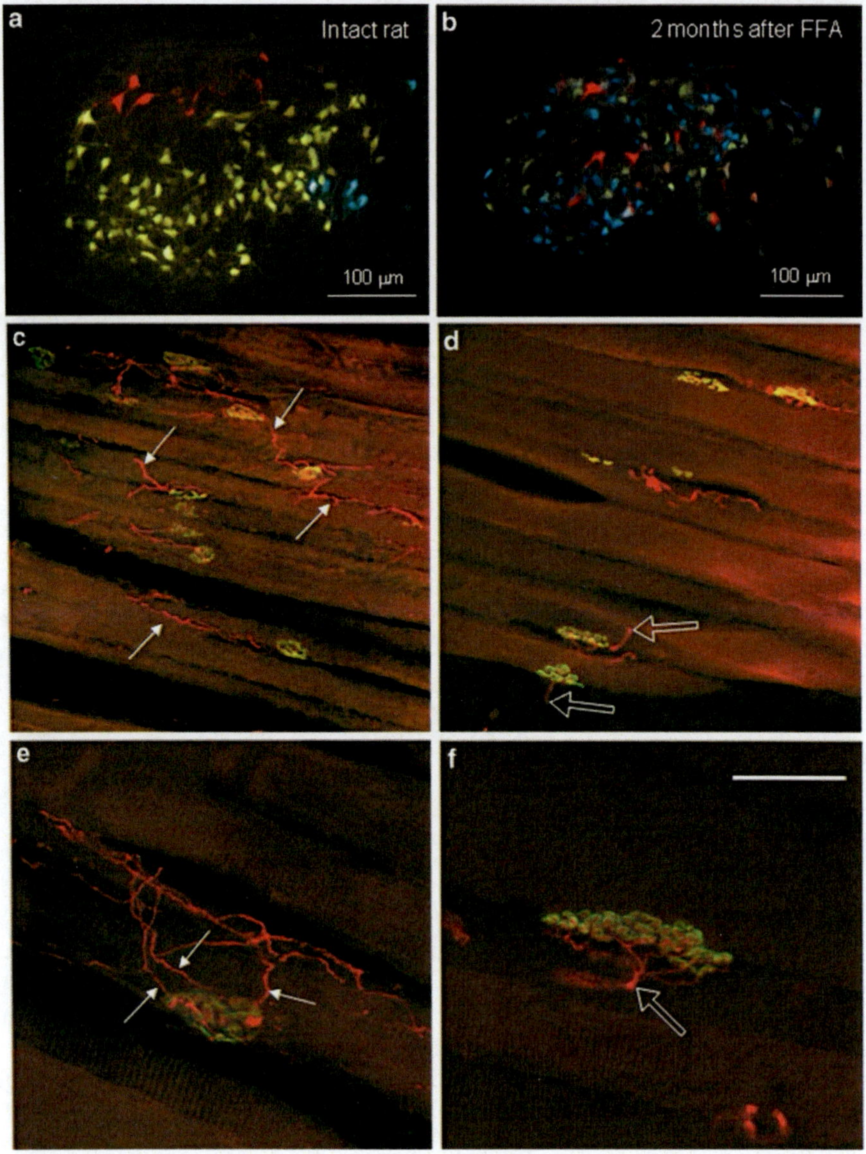

Fig. 2.2 (**a–d**) Myotopic organization of the facial nucleus and collateral axonal branching as estimated by the pattern of retrograde labeling. In intact animals, simultaneous application of DiI (*red*), FG (*yellow*), and FB (*blue*) to the zygomatic, buccal, and mandibular nerve branches, respectively, labels distinct subnuclei with no overlap (**a**). Two months after transection and suture of the facial nerve, the myotopic organization is lost irrespective whether the animals received ES (**b**). Adopted from Skouras et al. 2009. Superimposed

through the posterior auricular and cervical branches, respectively, branches that were neither transected nor labeled (data not shown, see Dohm et al. 2000; Streppel et al. 2002; Tomov et al. 2002).

2.1.1.14
Counting of Retrogradely Labeled Motoneuronal Perikarya

Employing the fractionator principle (Gundersen 1986), all retrogradely labeled motoneurons with visible cell nucleus in the 50-μm-thick sections were counted in every third section through the facial nucleus on the operated and on the unoperated side.

Eight weeks after unilateral FFA or entubulation and another 10 days after triple retrograde labeling, three major changes were detected. First, the myotopic organization into subnuclei was no longer observed, i.e., all retrogradely labeled motoneurons were scattered throughout the facial nucleus (Fig. 2.2b). Second, as a rule there were always more retrogradely labeled motoneuronal cell somata than in unoperated animals (Table 2.2). The reason for this was the postoperative hyperinnervation of targets (Angelov et al. 1996; Streppel et al. 1998), i.e., labeling of motoneurons which, under normal conditions, do not send axons to the three facial rami under study. After transection of the facial nerve, however, these motoneurons developed collateral axonal branches which adjoined the "wrong" rami and thus reached sites of tracer application. Third, numerous double-labeled motoneurons occurred after the lesion which demonstrated that twin axons projected into more than one branch of the facial nerve (Dohm et al. 2000; Streppel et al. 2002; Tomov et al. 2002). All double- and single-labeled motoneurons were counted.

Fig. 2.2 (continued) stacks of confocal images of endplates in reinnervated LLS muscles of a SD rat with normal vision (**c**, **e**) and a blind SD/RCS rat (**d**, **f**) visualized by staining of the motor endplates with Alexa Fluor 488 α-bungarotoxin (green fluorescence) and immunostaining of the intramuscular axons for neuronal class III β-tubulin (Cy3 red fluorescence) The images shown in panel **c** and **d** are taken at low magnification to reveal the pattern of innervation. Note that abundant intramuscular axonal branches are seen among endplates in panel **c** (*arrows*) but not in panel **d**. Also, the diameters of the muscle fibers seen in **c** are apparently smaller than those seen in **d**. Panels **e** and **f** show examples of a polyinnervated and a monoinnervated endplate, respectively. Three axonal branches (*arrows* in **e**) reach the boundaries of the polyinnervated endplate delineated by the alpha-bungarotoxin staining. By contrast, the monoinnervated endplate is reached by a single axon (*empty arrow* in **f**) with several preterminal rami. In both examples, the whole endplates are within the stack of confocal images. Scale bar shown in **f** indicates 125 μm for **c**, **d** and 40 μm for **e**, **f**. Adopted from Guntinas-Lichius et al. (2005)

2.1.1.15

Quantitative Estimates of the Total Number of Neurons Projecting Through the Zygomatic, Buccal, and Marginal Mandibular Branches of the Facial Nerve

Following FFA, the number of motoneurons projecting through the zygomatic, buccal, and marginal mandibular branch was increased by about 1.9-, 1.3-, and 3.9-fold, respectively, as compared with intact animals (last three columns in Table 2.2, compare groups *Intact* and *FFA*). Application of collagen to the transected and entubulated nerve did not result in substantial differences as compared with FFA (Table 2.2, group *Collagen*). By contrast, application of neutralizing antibodies (groups NGF through CNTF in Table 2.2) efficiently reduced the overall postlesional axonal branching as indicated by lower numbers of motoneurons labeled through the zygomatic (57–81% of control FFA value), the buccal (63–86%), and marginal mandibular branch (37–47%).

2.1.1.16

Quantitative Estimates of the Number of Single-Labeled (DiI-Only) and Double-Labeled (DiI + FG and DiI + FB) Motoneurons Projecting Through the Zygomatic Branch

The zygomatic branch was selected to evaluate the degree of axonal misdirection because the number of motoneurons that project(ed) through it before and after surgery was relatively small which allowed for sophisticated quantitative analysis in reasonable time. Thus, the distribution of the tracer DiI was of special interest and we carefully differentiated the proportions of motoneurons labeled by DiI_{only}, by DiI + FG, or by DiI + FB. The relative sizes of these three populations of neurons (columns 2–4 in Table 2.2) are indicated as percentages of the total population of DiI-labeled cells (column 5 in Table 2.2) below the absolute mean numbers. In intact rats the zygomatic branch consisted entirely (100%) of unbranched axons as indicated by the presence of DiI_{only}, but not of DiI + FG or DiI + FB cells. After FFA, DiI-labeled motoneurons projected through the three branches in roughly equal proportions (around 30%) indicating unselective, random growth of reinnervating axons into the three branches (Table 2.2). Application of antibodies not only reduced the total number of motoneurons labeled by DiI application to the transected zygomatic branch (see above) but also dramatically increased specificity of reinnervation: the numbers of double-projecting and thus double-labeled neurons (DiI + FG and DiI + FB) were significantly reduced. Accordingly, the portion of single-projecting and thus single-labeled cells that were labeled by DiI_{only} (64–79%) was significantly higher when compared with that after FFA (Table 2.2).

2.1.1.17

Analysis of Target Muscle Reinnervation

The reduction in collateral axonal branching at the lesion site failed to promote restoration of coordinated muscle activity. Considering the possibility that recovery of failure might be due to polyinnervation of motor endplates, we looked for parallelism between the portion of polyinnervated motor endplates and vibrissal motor performance in animals with poor recovery of whisking (SD rats with normal vision) and in rats with perfect recovery of vibrissae motor performance (RCS/SD blind rats). As a representative of the external vibrissal muscles we selected the m. levator labii superiors (LLS; Fig. 2.1c).

Two months after unilateral FFA, the SD and the SD/RCS animals were fixed by perfusion (see above). Under a surgical microscope, LLS on both operated and unoperated side of the face was dissected. Longitudinal sections (30 μm thick) were cut on a cryostat and mounted on SUPERFROST/Plus slides (Carl Roth, Karlsruhe, Germany). To visualize intramuscular axons and motor endplates, every third section through the muscle was stained with antineuronal class III β-tubulin and alpha-bungarotoxin.

Cryosections were immunostained with the polyclonal antineuronal class III β-tubulin (Covance, Richmond, CA, USA, No. PRB-435P) and anti-rabbit IgG Cy3 conjugate. Specificity controls (omission of the primary antibody or of the secondary biotinylated antibody) yielded blank sections. To visualize the motor endplates in the same sections, we stained the postsynaptic nicotinic acetylcholine receptors (nicotinic AchRs) with Alexa Fluor 488-conjugated α-bungarotoxin (Molecular Probes, B-13422; dilution 1:1,000 for 2 h at room temperature). Sections were observed with a Zeiss Axioskop 50 epifluorescence microscope through the "rhodamine" filter (excitation BP 546/12, beamsplitter FT 580, emission LP 590) and the "fluorescein" filter (excitation BP 450–490, beamsplitter FT510, emission LP 520).

Quality of endplate reinnervation was evaluated by a simple and straightforward criterion: number of axonal branches (identified by beta-tubulin staining, Fig. 2.2c–f) that enter or, in some cases, possibly leave the boundaries of individual endplates (identified by acetylcholine receptor staining with alpha-bungarotoxin, Fig. 2.2c–f). Entries by preterminal branches of one axon were counted as single events (Fig. 2.2f). According to this criterion, the endplates were identified as "monoinnervated" (one axon), "polyinnervated" (two or more axons), or denervated (no visible axonal associated with the receptor staining). The designation "polyinnervated" endplates is used to indicate similarity to a morphological abnormality in adult skeletal muscle of mammals observed in pathological conditions such as nerve damage or intoxication, which cause axonal branching, either collateral (at nodes of Ranvier) or terminal (from endplate terminals), or both (Shawe 1954;. Brown et al. 1981; Rich and Lichtman 1989), and polyneuronal innervation of individual muscle fibers. Individual endplates were identified by the alpha-bungarotoxin staining, after which the number of axons crossing the boundaries of the receptor staining was determined by focusing through the depth of the section. The frequencies of monoinnervated, polyinnervated, and noninnervated endplates were expressed in percentage of the total population.

Our qualitative examinations of longitudinal sections revealed two important differences between the two groups of animals: the incidence of intramuscular axonal branches was higher and diameters of muscle fibers were apparently smaller in muscles of SD rats as compared with SD/RCS animals (Fig. 2.2c, d).

The parallel assessment of vibrissal function (Fig. 2.3a) and of quality of endplate innervation (Fig. 2.3b) also revealed significant differences between the two groups of animals. Vibrissal movements were largely impaired in SD rats as indicated by the small values of the three most important functional variables

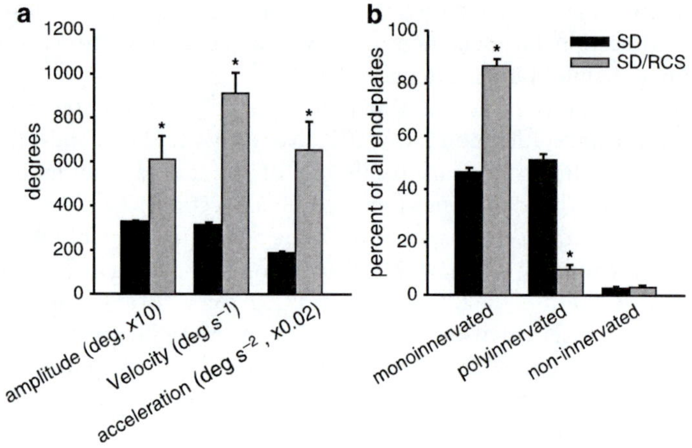

Fig. 2.3 (**a**, **b**) Results of quantitative assessment of vibrissae motor function (**a**) and evaluation of polyinnervation of muscle fibers in LLS muscles (**b**) in SD rats with normal vision (*black bars*) and blind SD/RCS rats (*gray bars*, $n = 6$ for both groups). The animals were studied 2 months after FFA anastomosis. The values for velocity and acceleration shown in **a** are for the protraction phase of the vibrissal movements (see text for further details). Endplates in the LLS muscles were classified as monoinnervated, polyinnervated, or noninnervated according to the number of beta-tubulin-immunoreactive axons that crossed the boundaries of the endplate delineated by staining for acetylcholine receptors (one, two, or more, and 0 axons for the three categories, respectively). The average number of endplates examined in every third section from the muscles was 556 ± 52 and 547 ± 79 per animal in the SD and SD/RCS group of rats, respectively. Similar numbers of endplates were examined in the contralateral intact LLS muscles of the same animals. In this case, no polyinnervated or noninnervated endplates were observed. The values shown in the graph are mean values + SEM. *Asterisks* indicate significant differences between the group mean values ($p < 0.05$, two-sided t test for independent sample). Adopted from Guntinas-Lichius et al. (2005)

estimated (Fig. 2.3a, *black bars*). By contrast, the values in the SD/RCS group were comparable with values of intact animals indicating an extraordinary high degree of recovery (Fig. 2.3a, *gray bars*, Table 2.2, Tomov et al. 2002). With regard to the differences in functional performance, the finding that muscle fibers in SD rats appear atrophic (see above) is not surprising. Poor function in the SD animals correlated with high percentage of polyinnervated endplates (>50%) in the LLS muscles (Fig. 2.3b, black bars). In the well-performing SD/RCS rats, the fraction of polyinnervated endplates was small (10%, Fig. 2.3b, *gray bars*). Analyses of endplates were also performed in sections from the intact contralateral LLS muscles of the same animals. All endplates observed in these samples were classified as monoinnervated (see Fig. 2.2 for the sample size for different groups of muscles and animals).

2.1.2
Local Stabilization of Microtubule Assembly Improved Recovery of Facial Nerve Function After Repair

To elucidate the relationship between collateral axonal branching and recovery of function, Peeva et al. (2006) analyzed the expression of cytoskeletal proteins after axotomy (Tetzlaff et al. 1988a) and determined the immunoreactivity for f-actin and neuronal class III β-tubulin, two cytoskeletal components responsible for cytomechanical forces in the leading edge of elongating axons (Challacombe et al. 1996). Using a stereological approach, Peeva et al. (2006) also estimated axonal growth cone densities in order to correlate degrees of axonal sprouting with protein expression levels. The results of this study provided experimental evidence for a possible relationship between functional outcome of facial nerve repair, on one side, and amounts of neurite growth-related proteins and numbers of outgrowing sprouts, on the other side. Better restoration of vibrissae motor performance appeared to be associated with a more vigorous early regenerative response. However, were these large amounts of tubulin due to an increased synthesis, to a slower transport, or to delayed turnover of tubulin? To prove perturbance of the microtubule assembly (polymerization and depolymerization), we applied established pharmacological agents (nocodazole, vinblastine, taxol) locally to the tips of the regrowing axons.

2.1.2.1
Animal Groups and Overview of Evaluation Methods

A total of 102 rats were divided into ten groups (1–10) (Table 2.3). Group 1 consisted of 12 intact control animals and groups 2–10 of surgically treated rats.

Two months after surgery on the buccal branch of the facial nerve (BBFN) and various treatments, the recovery of vibrissal motor performance was assessed in all rats using VBMA of explorative whisking. Thereafter in the experiments with altered local microenvironment at the lesion site (groups 2–7, all of 12 rats), half of

Table 2.3 Experimental design chart depicting animal grouping, treatments and parameters investigated

Group of animals	Video-based motion analysis of vibrissae motor performance (VBMA)	Degree of collateral axonal branching as estimated by double retrograde labeling	Amount of tubulin in BBFN as estimated by intensity of fluorescence	Pattern of motor endplate reinnervation in the LLS muscle
1. Intact animals (12 rats)	12	6	6	6
2. Surgery for buccal–buccal anastomosis (BBA; 12 rats)	12	6	6	6
3. Tube with 0.1 M phosphate buffer pH 7.4 over the transected BBFN (12 rats)	12	6	6	6
4. Tube with collagen type I over the transected BBFN (12 rats)	12	6	6	6
5. Tube with 100 μg/ml nocodazole over the transected BBFN (12 rats)	12	6	6	6
6. Tube with 20 μg/ml vinblastine over the transected BBFN (12 rats)	12	6	6	6
7. Tube with 10 μg/ml taxol over the transected BBFN (12 rats)	12	6	6	6
8. BBA + injection of nocodazole (100 μg/ml) into the whisker pad muscles (six rats)	6	–	–	6
9. BBA + injection of vinblastine (20 μg/ml) into the whisker pad muscles (six rats)	6	–	–	6
10. BBA + injection of taxol (10 μg/ml) into the whisker pad muscles (six rats)	6	–	–	6

In groups 1–7, half of the animals that underwent video-based motion analysis (VBMA) were subsequently used to estimate the degree of collateral axonal branching and the other half to determine the amount of tubulin in the regrowing buccal branch of the facial nerve (BBFN) and the pattern of motor endplate reinnervation. In groups 8–10, the animals that were subjected to VBMA were thereafter used only for establishing the pattern of motor endplates reinnervation the animals of each group were used to establish the degree of posttransectional collateral axonal branching using retrograde labeling with two different fluorescent dyes applied simultaneously to both peripheral divisions of BBFN. The other half of the rats were used to study possible alterations in the amount of microtubules in the regrowing BBFN (quantitative immunohistochemical analysis of the expression of neuronal class III β-tubulin) and quantitative aspects of the target muscle reinnervation by means of immunocytochemistry for neuronal class III β-tubulin and histochemistry with alpha bungarotoxin. In groups 8–10 (each consisting of six rats), only the vibrissae motor performance and reinnervation pattern were analyzed.

2.1.2.2

Transection and Suture of the Buccal Branch of the Facial Nerve (Buccal–Buccal Anastomosis)

Transection and suture of the buccal branch of the facial nerve (buccal–buccal anastomosis, BBA) was performed only in animals of major groups B. The right BBFN was transected and immediately reconnected to the distal stump. Since the subsequent evaluations included analysis of the vibrissae motor performance, we had to eliminate any additional motor nerve supply to the whisker pad muscles (Semba and Egger 1986). Therefore, the transection of the buccal branch was always accompanied by transection and proximal ligature (to prevent regeneration) of the marginal mandibular branch of the facial nerve (Fig. 2.4a).

2.1.2.3

Entubulation of the Buccal Branch of the Facial Nerve

Under anesthesia, the right BBFN in groups C–G was transected and both stumps were inserted into a silicone precision tube (Fig. 2.4b). The space between the proximal and distal nerve stumps was filled with 0.1 M phosphate buffered saline pH 7.4 (group 3), collagen type I (group 4), or with collagen gel containing nocodazole (group 5), taxol (group 6), and vinblastine (group 7).

2.1.2.4

Treatments

For the *nocodazole* treatment, each silicon tube was prefilled with a gel consisting of 66 μl collagen type I (Serva, Cat. No. 47254) and 33 μl of solution containing 10.0 μg of nocodazole (Sigma, Cat. No. M1404). Alternatively, the 800 μl collagen was obtained from 5 ml collagen stock solution. The three-dimensional collagen gel (Guidry and Grinnell 1987; Mauch et al. 1988) with the nocodazole solution was left to polymerize for 2 h at 37°C. After mixing the 400 μl nocodazole solution with 800 μl collagen solution, the therapeutic concentration of 100 μg/ml was reached (Chuckowree and Vickers 2003).

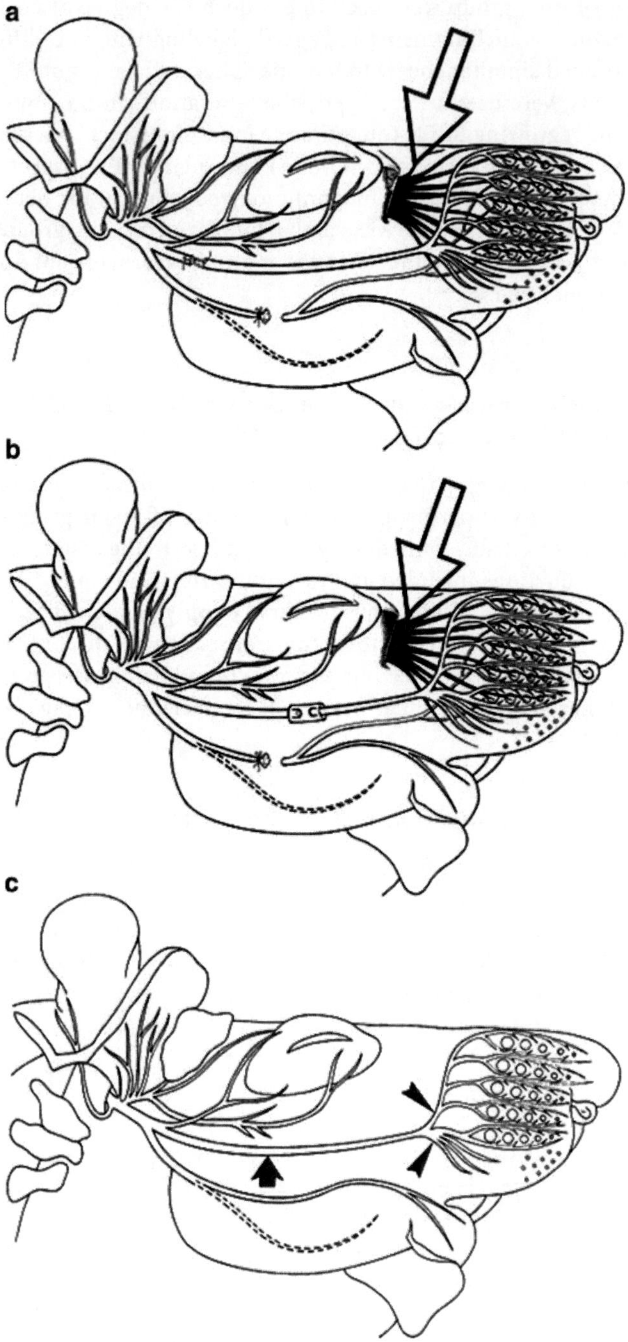

Fig. 2.4 (a–c) Surgical operations on the rat buccal branch of the facial nerve (BBFN). Schematic drawings illustrating the sites of transection and suture (**a**) and entubulation (**b**)

For the *vinblastine* treatment, the silicon tube was prefilled with a gel consisting of 67 μl collagen type I and 33 μl of 12 nM solution of vinblastine (Sigma, V1377). After mixing the 400 μl vinblastine solution with 800 μl collagen (see above), the therapeutic concentration of 4 nM or 20 μg/ml was reached (Challacombe et al. 1997).

For the *taxol* treatment, the silicon tube was prefilled with a gel consisting of 67 μl collagen type I and 33 μl of solution containing 1 μg paclitaxel (Sigma, T1912). After mixing the 400 μl paclitaxel solution with 800 μl collagen, the therapeutic concentration of 10 μg/ml was reached (Chuckowree and Vickers 2003).

2.1.2.5 Analysis of Vibrissae Motor Performance

Two months after transection and entubulation of BBFN, denervation-induced changes were found in all values (except for the frequency) selected to determine the functional state (Table 2.4). The statistical analysis revealed that only the animals in group 7 (treatment with 10 μg/ml taxol) performed significantly better than those in group 4 (entubulation in collagen; Table 2.4).

2.1.2.6 Simultaneous Application of Two Fluorescent Tracers

One day after videotaping, half of the intact rats of group A and half of the surgically treated rats from groups 2–7 received an intraperitoneal injection of ketamin/xylazin. The BBFN was exposed, and the superior and inferior buccolabial nerves on the right side of the face were transected and labeled with crystals of Fluoro-Gold (FG) and DiI, respectively (Fig. 2.4c).

2.1.2.7 Tissue Preparation, Fluorescence Microscopy, Image Analysis, and Counts

Intact Rats

Application of Fluoro-Gold crystals to the superior and DiI crystals to the inferior buccolabial nerve yielded 1,536 ± 243 FG- and 134 ± 125 DiI-labeled motoneurons (mean ± SD, $n = 6$). All retrogradely labeled cells (total of 1,682 ± 338) were localized in the lateral facial subnucleus. Thereby, the FG-labeled cells were found in its ventral and the DiI-labeled perikarya in its dorsal portion. No double-labeled motoneurons were observed (Fig. 2.5a).

Fig. 2.4 (continued) of BBFN and of the transection and ligature of the marginal mandibular branch of the facial nerve. The cervical branch of the facial nerve is indicated by a dotted line. Adopted from Skouras et al. (2002). (c) Schematic drawing of the infratemporal portion of the rat facial nerve. *Large arrow* indicates the transection, suture and entubulation site of BBFN. Transection and tracer application sites in the superior and inferior buccolabial nerves are indicated by *arrowheads.* Adopted from Angelov et al. (1999). The infraorbital nerve (ION) is indicated by an *arrow*

Table 2.4 Recovery of vibrissae motor performance function after buccal nerve lesion in rats

Group of animals	Frequency (in Hz)	Angle at maximal protraction (in degrees)	Amplitude (in degrees)	Angular velocity during protraction (in degrees/s)
1. Intact animals (12 rats)	6.4 ± 1.1	71 ± 15	50 ± 15	588 ± 276
2. Buccal–buccal anastomosis (BBA; 12 rats)	7.2 ± 0.8	98 ± 16	20 ± 6	326 ± 134
3. Tube with phosphate buffer (12 rats)	7.4 ± 0.6	95 ± 11	21 ± 4	404 ± 106
4. Silicon tube with collagen type I (12 rats)	7.5 ± 0.7	88 ± 18	21 ± 5	327 ± 108
5. Tube with 100 μg/ml nocodazole (12 rats)	7.4 ± 0.7	92 ± 14	22 ± 4	428 ± 116
6. Silicon tube 20 μg/ml vinblastine (12 rats)	7.2 ± 0.4	96 ± 17	20 ± 2	380 ± 187
7. Silicon tube with 10 μg/ml taxol (12 rats)	7.2 ± 0.9	84 ± 17	34 ± 8*	634 ± 164*
8. BBA with subsequent injection of nocodazole (100 μg/ml) into whisker pad muscles (six rats)	7.1 ± 1.7	71 ± 20	28 ± 7	400 ± 344
9. BBA with subsequent injection of vinblastine (20 μg/ml) into whisker pad muscles (six rats)	5.2 ± 0.9	90 ± 14	20 ± 9	340 ± 291
10. BBA with subsequent injection of taxol (10 μg/ml) into whisker pad muscles (six rats)	6.3 ± 1.1	86 ± 14	20 ± 15	216 ± 188

Shown are group mean values ± SD. Group mean values significantly different from the control group D (entubulation with collagen; ANOVA and post hoc Dunnett's test, $p < 0.05$) are indicated by asterisks. Values for intact rats are given as reference values and not included in the analysis

Operated Rats

Neuron labeling at 2 months after transection and/or entubulation of BBFN showed that all retrogradely labeled neurons were localized in the lateral facial subnucleus. However, due to a malfunctioning guidance of regrowing axons into wrong fascicles, the myotopic organization of this subnucleus into a ventral (for the superior buccolabial nerve) and a dorsal (for the inferior buccolabial nerve) portion was no longer evident (Fig. 2.5b–d). Accordingly, the number of motoneurons whose axons or axonal branches projected into the superior buccolabial nerve was lower than in the intact rats (91%). We determined that only about 28–56% of all neurons in the lateral facial subnucleus projected into the superior buccolabial nerve. On the contrary, the number of motoneurons whose axons projected into the inferior buccolabial nerve was increased in comparison with that in the intact rats: the motoneurons whose axons had regrown into the inferior buccolabial nerve comprised about 24–48% of all neurons in the lateral facial subnucleus. Compared with the values in intact rats (9%), there was a statistically significant decrease in the number of motoneurons projecting through the superior buccolabial nerve.

Another major difference to the unoperated animals was the presence of motoneurons containing both tracers. The only explanation for this may be that these double-labeled cells (approximately 21% of all motoneurons in the lateral facial

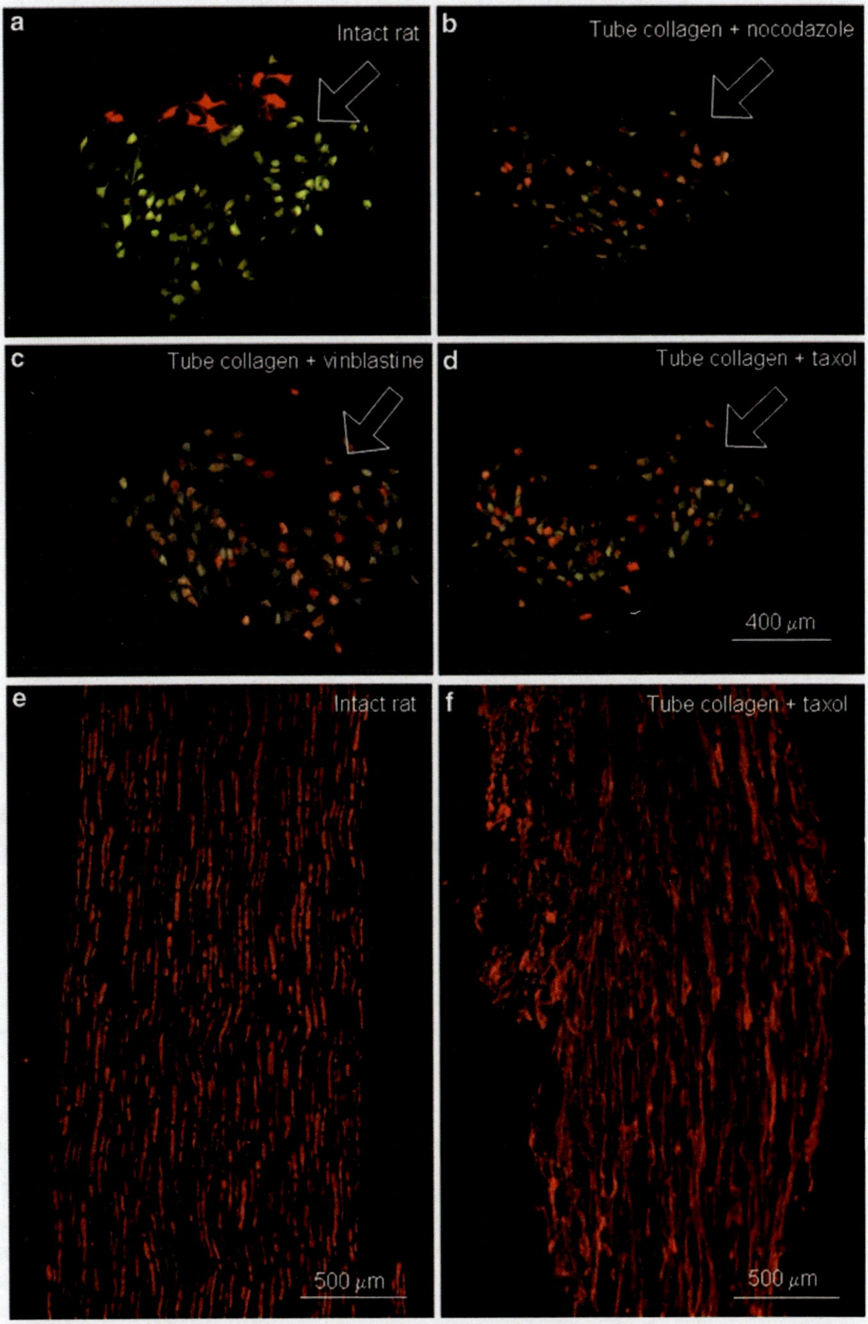

Fig. 2.5 (a–f) Facial nucleus after retrograde neuronal labeling in intact and operated rats. Photographs produced by double exposure. The dorsomedial border of the lateral facial

subnucleus) regrew several (but not one single) sprouts from each transected axon, which postoperatively projected simultaneously within the superior and inferior buccolabial nerves (cf. Shawe 1954). Since both rami were instilled with different tracers, the axonal branches took up and retrogradely transported two fluorescent dyes simultaneously. This in turn led to double labeling of their parent perikarya.

Two months after entubulation with collagen (group 4) we determined an index of axonal branching of 21%. None of the other entubulations (group 5 – 21%; group 6 – 21%; group 7 – 22%) had any significant influence on the projection patterns (ANOVA and post hoc Dunnett's test, $p < 0.05$). Thus, there was a complete lack of myotopic organization and a consistently elevated degree of axonal branching regardless of whether the animals were subjected to mere transection of the BBFN or to entubulation with varying agents.

2.1.2.8 Intensity of Tubulin Fluorescence

One day after videotaping, the rest of the animals from groups A–G were deeply anesthetized and transcardially perfused with 4% formaldehyde for 20 min. The masseter muscle with the transected buccal branch of the facial nerve (BBFN) on it was removed, post-fixed, cryoprotected, and cut in 20-μm-thick frozen sections (Streppel et al. 2002; Peeva et al. 2006). All sections used to compare the intensity of fluorescence between the various groups were incubated simultaneously using identical solutions.

Immunocytochemistry with 1:1,000 rabbit polyclonal antineuronal class III β-tubulin (Covance, Richmond, CA, USA) was performed as previously described.

For quantification of pixel brightness, images were captured with a slow scan CCD camera (Spot RT, Diagnostic Instruments) using the ×16 objective and the Image-Pro Plus Software Version 5.0 (Media Cybernetics, Inc., Silver Spring, MD, USA). Analysis with a threshold for tubulin that had been set at pixel gray value of 120 was performed as previously described (Peeva et al. 2006).

Reliable immunostaining with the selected antibodies as described above allowed us to readily identify Cy3-fluorescent microtubules in longitudinal sections

Fig. 2.5 (continued) subnucleus is indicated by an *arrow*. (**a**) Unlesioned lateral facial subnucleus with preserved myotopic organization of the motoneurons whose axons project into the superior buccolabial nerve (retrogradely labeled in *yellow* by Fluoro-Gold) and into the inferior buccolabial nerve (labeled in *red* by DiI). Whereas most FG-labeled motoneurons are localized in the ventral portion, those labeled with DiI are in the dorsal part of the subnucleus. (**b**)–(**d**): Lesioned lateral facial subnucleus after treatment of the transected BBFN with nocodazole (**b**), vinblastine (**c**), and taxol (**d**) and application of FG to the superior and DiI to the inferior buccolabial nerve. Note the complete lack of myotopic organization: the FG-labeled (*yellow*), DiI-labeled (*red*), and DiI + FG-labeled (*shades of pink and orange*) motoneurons are scattered throughout the whole lateral facial subnucleus. 50 μm vibratome sections. Immunocytochemical demonstration of neuronal class III β-tubulin in rat BBFN axons. Representative longitudinal sections from an intact nerve (**e**) and from a fragment proximal to the silicon tube, which had been filled with 10 μg/ml taxol in collagen 2 months beforehand (**f**). Adopted from Grosheva et al. 2008 through the intact and lesioned BBFN (Fig. 2.5e, f). At least six equidistant (each second) 20-μm-thick longitudinal sections through the BBFN of each animal were measured, yielding a mean of about 40 sections per group (each experimental group consisted of six rats).

The results on the intensity of fluorescence showed that there was no significant difference in the amount of tubulin (indirectly determined by the number of pixels at the defined gray value of 120) between the intact BBFN (group 1: 5,493 ± 3,418) and the BBFN at 2 months after its transection and suture (group 2: 5,384 ± 2,145). Similar values were determined after entubulation of transected BBFN in phosphate buffer (group 3: 2,814 ± 1,873), collagen type I (group 4: 4,275 ± 924), nocodazole (group 5: 3,666 ± 957), vinblastine (group 6: 5,049 ± 2,185), and taxol (group 7: 2,158 ± 1,706). No statistically significant differences between any of the groups were detected.

2.1.2.9
Degree of Motor Endplate Polyinnervation

Degree of motor endplate polyinnervation was investigated in the m. levator labii superioris (Fig. 2.1c) as described (Sect. 2.1.1). The frequencies of monoinnervated, polyinnervated, and noninnervated endplates were expressed as percentage of the total population. Although postlesional polyinnervation of the endplates has been claimed to be transient (Hennig and Dietrichs 1994), accumulating evidence suggests that it persists after establishment of nerve–muscle contacts (Esslen 1960; Mackinnon et al. 1991; Reynolds and Woolf 1992; Madison et al. 1999; Jergovic et al. 2001; Ijkema-Paassen et al. 2002; Grant et al. 2002; Choi and Raisman 2005). Our previous work indicates that it has a deleterious effect on recovery of facial motor function (Guntinas-Lichius et al. 2005).

In intact animals, all motor endplates were monoinnervated (Fig. 2.2d, f; Table 2.5). Two months after any transection and end-to-end suture or entubulation of the BBFN (groups B–F), about 30% of the motor endplates were polyinnervated, i.e., innervated by two or more axons (Fig. 2.2c, e; Table 2.5). The only procedure that significantly reduced the degree of polyinnervated endplates (12 ± 4%) was entubulation in 10 μg/ml taxol.

2.2
Efforts to Reduce Axonal Sprouting in Denervated Muscles

2.2.1
Direct Modification of Microtubule Dynamics in Reinnervated Muscles Failed to Reduce Terminal Axonal Sprouting

Our results described in Sect. 2.1.2 showed that stabilization of microtubules with 10 μg/ml taxol reduced intramuscular axonal sprouting and polyinnervation of

Table 2.5 Reinnervation pattern of the m. levator labii superioris (LLS)

Group of animals	Monoinnervated motor endplates (%)	Polyinnervated motor endplates (%)	Noninnervated motor endplates (%)	Total number of motor endplates examined
1. Intact animals	100 ± 0	0	0	1,521 ± 111
2. Buccal–buccal anastomosis	67 ± 10	29 ± 7	4 ± 3	1,763 ± 131
3. Tube with 0.1 M phosphate buffer pH 7.4	71 ± 6	26 ± 5	3 ± 1	1,868 ± 313
4. Tube with collagen	70 ± 7	24 ± 4	2 ± 1	1,650 ± 221
5. Tube with 100 μg/ml nocodazole	69 ± 12	27 ± 5	4 ± 3	1,426 ± 73
6. Tube with 20 μg/ml vinblastine	74 ± 9	23 ± 4	3 ± 1	1,482 ± 161
7. Tube with 10 μg/ml taxol	86 ± 9*	12 ± 4*	2 ± 1	1,362 ± 149
8. BBA with subsequent injection of nocodazole (100 μg/ml) into the whisker pad muscles	59 ± 14	37 ± 5	4 ± 3	1,722 ± 312
9. BBA with subsequent injection of vinblastine (20 μg/ml) into the whisker pad muscles	51 ± 16	45 ± 5	4 ± 1	1,560 ± 212
10. BBA with subsequent injection of taxol (10 μg/ml) into the whisker pad muscles	60 ± 15	36 ± 5	4 ± 2	1,716 ± 318

Motor endplates were classified as monoinnervated, polyinnervated, or noninnervated according to the number of beta-tubulin-immunoreactive axons that crossed the boundaries of the endplate. Shown are group mean values ± SD. Group mean values significantly different from the control group D (entubulation with collagen; ANOVA and post hoc Dunnett's test, $p < 0.05$) are indicated by asterisks. Values for intact rats are given as reference values and not included in the analysis

the motor endplates which was accompanied by improved restoration of function. This led to the question whether direct modification of microtubule dynamics in reinnervated muscles by local application of nocodazole, vinblastine, and taxol would reduce the terminal axonal sprouting and thus diminish the portion of polyinnervated motor endplates.

Earlier own experience has shown that, after BBA, the first regrowing axons bridge the gap between transection site and target muscles (about 12 mm long) for 4 days (Angelov et al. 1999). This is why starting at 5 days after BBA, rats received 50 μl injections of nocodazole (10 μg/100 μl), vinblastine (2 μg/100 μl), and taxol (1 μg/100 μl) into muscles of the whisker pad once in a week.

Analysis of vibrissae motor performance and analysis of target muscle reinnervation were performed as described. Contrary to our expectation that direct modification of microtubule dynamics in reinnervated muscles would reduce the terminal axonal sprouting and thus diminish the portion of polyinnervated motor endplates, we found an average of 35–45% polyinnervated muscle fibers after local application of nocodazole, vinblastine, and taxol and poor recovery of vibrissal motor performance (Tables 2.4 and 2.5).

2.2.2
Intraoperative Electrical Stimulation Prior to Reconstructive Surgery Did Not Improve Recovery of Function

Recently, a novel clinically feasible approach to enhance peripheral nerve regeneration after femoral nerve lesion in rats was suggested (Al-Majed et al. 2000; Brushart et al. 2005; Geremia et al. 2007; Gordon et al 2007). Briefly, low-frequency intraoperative electrical stimulation (IOES; 1 h, 20 Hz) was delivered to the proximal nerve stump of the severed nerve prior to surgical reconstruction. Stimulation led to depolarization of the motoneuron perikarya and a significant shortening of the period of asynchronous, "staggered" axonal regrowth (Al-Majed et al. 2000; Brushart et al. 2002). These beneficial effects were associated with a faster and enhanced upregulation of brain-derived neurotrophic factor (BDNF) and its tyrosine kinase B (TrkB) receptor in motoneurons (Al-Majed et al. 2004; English et al. 2007); in addition, TrkB-dependent expression of the HNK-1 (human natural killer cell antigen-1) glycoepitope was increased in the quadriceps branch of the femoral nerve (Eberhardt et al. 2006). Brief electrical stimulation after sciatic nerve injury also promoted axonal regeneration and attenuates facilitation of spinal motor responses (Vivó et al. 2008). Could this therapy be also successful after facial nerve lesion?

2.2.2.1
Animal Groups and Overview of Experiments

Forty-eight rats were distributed in three groups each of 16 animals. Group 1 consisted of intact rats and groups 2 and 3 consisted of experimental rats that were subjected to unilateral transection and suture of the right facial nerve (FFA; Fig. 2.1a). Rats in group 2 (sham stimulated, SS, Fig. 2.6b) had electrodes placed on the proximal stump after nerve transection, but no electric current was applied. Rats in group 3 were subjected to electrical stimulation of the facial nerve immediately after facial nerve transection but prior to end-to-end suture (Fig. 2.6a).

In groups 2 (SS) and 3 (ES), vibrissal motor performance was evaluated in the same animals at 1, 2, 3, and 4 months after surgery. The degree of axonal branching and pattern of motor endplate reinnervation were determined at the end of the experiment (i.e., 4 months). Vibrissal motor performance during explorative whisking was analyzed using VBMA. Following the last functional analysis 4 months post-surgery, half ($n = 8$) of the animals in all groups were used to investigate the degree of collateral axonal branching by using triple

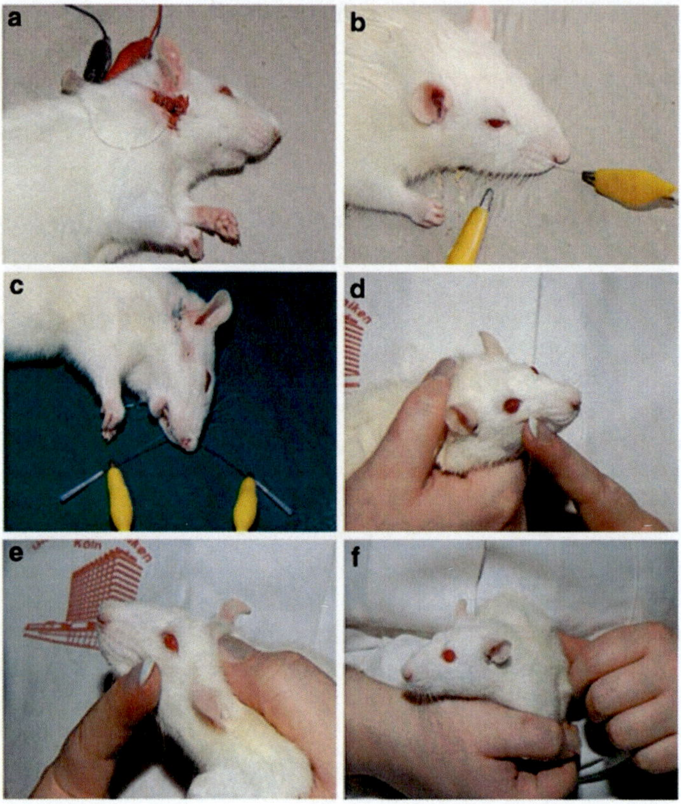

Fig. 2.6 (a–f) Intraoperative electrical stimulation of the proximal stump of the transected right facial nerve (a). Adopted from Skouras et al. (2009). (b) Sham stimulation of rats: acupuncture needle electrodes were inserted, but no current was applied to the electrodes (c) Postoperative electrical stimulation of the vibrissal muscles. Adopted from Sinis et al. 2009. (d) Manual mechanical stimulation of the right, i.e., ipsilateral to the nerve transection and suture (FFA) vibrissae and whisker pad muscles. (e) Manual mechanical stimulation of the left, i.e., contralateral to FFA vibrissae and whisker pad muscles. (f) Handling of the animals. Adopted from Angelov et al. (2007)

retrograde neuronal labeling. The remaining rats ($n = 8$) were used to determine the proportion of monoinnervated and polyinnervated motor endplates in the ipsilateral levator labii superioris muscle using immunocytochemistry for neuronal class III β-tubulin and histochemistry with alpha-bungarotoxin (see below).

2.2.2.2

Intraoperative Electrical Stimulation

IOES was performed as described by Ahlborn et al. (2007). The right facial nerve was exposed and a Teflon-coated stainless steel wire (50 μm in diameter, bared of insulation at its tip) was twisted to form a loop around the nerve stump. A second electrode, used as an anode, was fixed to a muscle close to the nerve. In all electrically stimulated rats ($n = 16$), the threshold voltage required to elicit visible contractions of the whisker pad muscles was determined by applying square 0.1 ms pulses at 20 Hz at varying voltage intensities using a pulse generator (Master-8, A.M.P.I., Jerusalem, Israel). Immediately thereafter, the nerve stump was transected with fine scissors about 2 mm distally from the electrode. The proximal nerve stump was then stimulated for 1 h by applying square 0.1 ms pulses at 20 Hz using amplitudes three times above threshold levels (typically 3–4 V; Fig. 2.6a). Thereafter, the electrodes were removed, and the ends of the nerve were sutured with single epineural 11-0 nylon stitches (Ethicon, Norderstedt, Germany). Control sham-stimulated rats were treated similarly to rats subjected to ES except that no current was applied to the electrodes.

2.2.2.3

Analysis of Vibrissae Motor Performance

Analysis of vibrissae motor performance was performed as described (Sect. 2.1.1). At 1–4 months after nerve transection and sham stimulation (Groups 2a–d, Table 2.6), vibrissal motion was poor as compared with intact animals. The mean amplitude and angular velocity of vibrissal movements were reduced to 25–33% and 9–12% of control values, respectively (fourth and fifth column in Table 2.6), and the angle of maximal protraction was increased by more than 50% (third column in Table 2.6).

IOES (Groups 3a–d, Table 2.6) improved neither the amplitude of vibrissal movements nor the angle of maximal protraction as compared with sham stimulation. The angular velocity during protraction was, however, significantly higher (+52–92%) in electrically stimulated rats than in sham-stimulated rats at 1–3, but not at 4 months after surgery. We conclude that IOES does not improve the final outcome of FFA but causes some limited functional advantage in the first three postoperative months.

2.2.2.4

Application of Three Fluorescence Tracers, Fluorescence Microscopy, and Counting

In intact animals, motoneurons with axons entering the zygomatic, buccal, and marginal mandibular ramus were localized in the dorsal, lateral, and intermediate facial subnuclei, respectively (Semba and Egger 1986). No double- or triple-labeled motoneurons were observed because intact motoneurons send only one unbranched axon to one of the facialis rami (Fig. 2.2a). Thus, the index of axonal branching in the facial nerve trunk of intact animals, calculated from the zygomatic motoneurons (sum of the percentages in the third and fourth column of Table 2.7), was 0%.

Table 2.6 Motor recovery after facial nerve lesion and intraoperative electrical stimulation (IOES)

Group of animals	Frequency (in Hz)	Angle at maximal protraction (in degrees)	Amplitude (in degrees)	Angular velocity during protraction (in degrees/s)
1. Intact	7.0 ± 0.8	62.0 ± 13	57 ± 13	1,238 ± 503
2. FFA + SS				
2a. 1 month after FFA	5.8 ± 0.8	102 ± 8.9#	15 ± 2.4#	107 ± 26#
2b. 2 months after FFA-	6.3 ± 0.5	91 ± 12#	19 ± 6.1#	135 ± 54#
2c. 3 months after FFA-	6.1 ± 0.8	101 ± 15#	16 ± 4.2#	154 ± 38#
2d. 4 months after FFA	6.4 ± 0.9	111 ± 11#	14 ± 3.1#	114 ± 22#
3. FFA + IOES				
3a. 1 month after FFA	6.1 ± 1.1	99 ± 10#	13 ± 4.2#	205 ± 114*#
3b. 2 months after FFA	6.3 ± 1.2	80 ± 13#	17 ± 9.2#	235 ± 74*#
3c. 3 months after FFA	6.6 ± 1.1	93 ± 38	18 ± 13#	235 ± 53*#
3d. 4 months after FFA	6.4 ± 0.9	96 ± 11#	11 ± 6.1#	136 ± 48#

Biometrics of vibrissae motor performance in intact rats (*Intact*), in rats subjected to transection and suture of the facial nerve (FFA) plus 1 h intraoperative sham stimulation (FFA + SS) or 1 h intraoperative electrical stimulation (FFA + IOES) of the proximal stump of the transected facial nerve. All groups consisted of 16 rats. Shown are group mean values ± SD. Asterisks indicate differences between mean values of electrically (ES) and sham-stimulated (SS) animals (group 2a–d) at given postoperative time-point (ANOVA for repeated measurements and post hoc Tukey's test, $p < 0.05$)

#Indicate differences in mean values between intact rats and the experimental groups (ANOVA for repeated measurements and post hoc Tukey's test, $p < 0.05$) and show the expected decline in function

Four months after facial nerve cut and suture with sham stimulation (FFA-SS, Group 2), no myotopic organization into subnuclei was observed, i.e., all retrogradely labeled motoneurons were scattered throughout the facial nucleus (Fig. 2.2b). The same labeling pattern was observed when IOES was applied for 1 h prior to end-to-end suture (FFA-ES, Group 3). The lack of myotopy in all groups presumably reflects poor axonal pathfinding and misdirection of the regrowing axons after transection and suture of the facial nerve, a robust feature which could not be overridden by IOES. No fascicular orientation in the zygomatic, buccal, or marginal mandibular branches occurred.

Double and triple labeling was also commonly observed (Fig. 2.2b) and is explained by multiple axonal branches originating from individual perikarya (Shawe 1954) which grow simultaneously into different rami (i.e., zygomaticus, buccalis, and/or marginalis mandibulae); such sprouts retrogradely transported the different fluorescent dyes to their parent motoneurons in the facial nucleus.

Collateral axonal branching after transection of the facial nerve dramatically affected the fiber composition of the different nerve rami (Table 2.7). In intact rats (group 1), the total number of single-labeled (i.e., DiI-only + FG-only + FB-only) cells in the facial nucleus was 2,184 ± 242 but was increased to 3,622 ± 672 at 4 months after FFA plus sham stimulation (group 2) and to 3,233 ± 1,269 at 4 months after FFA + IOES (group 3). The high numbers of single-labeled motoneurons in the

Table 2.7 Degree of collateral axonal branching after facial nerve injury and intraoperative electrical stimulation (IOES)

Group of animals	Neurons projecting only into the zygomatic nerve (DiI-only)	Neurons projecting into the zygomatic and buccal nerves (DiI + FG)	Neurons projecting into the zygomatic and marginal mandibular nerves (DiI + FB)	All DiI labeled neurons projecting into the zygomatic nerve (DiI, DiI + FG, DiI + FB)	Neurons projecting only into the buccal nerve (FG-only)	Neurons projecting only into the mandibular nerve (FB-only)	Total number of single-labeled neurons (DiI-only + FG-only + FB-only)
1. Intact[1]	364 ± 47	–	–	364 ± 47	1441 ± 101	379 ± 94	2184 ± 242
	100%	0%	0%	100%			
2. 4 months	222 ± 35#	240 ± 29#	258 ± 56#	720 ± 120#	1911 ± 281#	1489 ± 356#	3622 ± 672#
FFA + SS	31%	33%	36%	100%			
3. 4 months	278 ± 114	161 ± 66#	181 ± 66#	619 ± 182#	1440 ± 709	1515 ± 446#	3233 ± 1,269#
FFA + IOES	45%	26%	29%	100%			

Number of motoneurons with axons in the zygomatic, buccal, or marginal mandibular branches of the facial nerve in intact rats (Intact), in rats subjected to transection and suture of the facial nerve (FFA) and 1 h intra-operative sham stimulation (FFA + SS) or 1 h electrical stimulation (FFA + IOES) of the proximal stump of the transected facial nerve. Animals were studied 10 days after triple retrograde labeling. At least eight animals were studied per group. Shown are group mean values ± SD. There were no significant differences between the stimulated group (3) and the nonstimulated group (2) (ANOVA, $p > 0.05$)

#Indicate significant differences (ANOVA, $p > 0.05$) between intact animals and the experimental group and show the expected increases in numbers of labeled neurons. The percentage values below the absolute numbers in columns 2–5 indicate the portions of motoneurons projecting through the zygomatic nerve with branched (DiI + FG or DiI + FB, column 3 and 4) and unbranched axons (DiI-only, column 2)

[1]Values have been adopted from Table 2.2

experimental groups following labeling of the buccal (FG-only) and marginal mandibular (FB-only), but not the zygomatic, branches (Table 2.7) indicate that other axons must have sprouted and entered these rami. Axons from the two other branches of the facial nerve (posterior auricular and cervical), while not labeled in intact animals, must presumably have sprouted into the zygomatic, buccal, and marginal mandibular rami (Table 2.7). However, the elevation revealed by retrograde tracing does not exceed the normal physiological number of neurons in the rat facial nucleus shown by other techniques [neuron-specific enolase (NSE) immunostaining: 4,066 ± 508 (Angelov et al. 1994); Nissl-staining: 3,835 ± 537 (Guntinas-Lichius et al. 1993); and retrograde labeling with horseradish peroxidase (Angelov et al. 1993; Streppel et al. 1998)]. Similar values are also seen at 1 (4,152 ± 166 facial neurons) and 8 weeks after facial nerve transection and suture (3,753 ± 273 facial neurons; Angelov et al. 1994). On the other hand, cell death after facial nerve axotomy occurs only if (1) rats are newborn (Umemiya et al. 1993; Clatterbuck et al. 1994; Rossiter et al. 1996; Moran and Graeber 2004), (2) the axotomy is followed by resection of approximately 1 cm nerve length which causes permanent deprivation from the target (Tetzlaff et al. 1988a, b), or (3) the facial nerve axotomy was performed in mice rather than in rats (de Bilbao and Dubois-Dauphin 1996; Deckwerth et al. 1996; Raivich et al. 1998, 1999; Moran et al. 2001). Taken together, the data indicate that motoneuron neurogenesis does not occur after facial nerve injury.

Retrograde tracing did not reveal any changes in the total numbers of single-labeled motorneurons (i.e., DiI-only + FG-only + FB-only) in the facial nucleus among the experimental groups (i.e., SS, ES; Table 2.7). The total number of single-labeled cells was 2,184 ± 242 in intact rats (group 1), 3,622 ± 672 after sham stimulation (FFA-SS, group 2), and 3,233 ± 1,269 following ES (FFA + IOES, group 3). Furthermore, the index of axonal branching was 55–69% (sum of the percentages of DiI + FG and DiI + FB retrogradely labeled perikarya in the third and fourth column of Table 2.7). In summary, the complete lack of myotopic organization, and a consistently elevated degree of axonal branching, suggested that IOES did not influence axonal projection patterns.

Pattern of target muscle reinnervation was estimated as described (Sect. 2.1.1).

In intact animals, all motor endplates were monoinnervated (Fig. 2.2d, f). After facial nerve injury, the proportion of polyinnervated motor endplates was 51 ± 10% after sham stimulation (FFA-SS, group 2) and 42 ± 4% after IOES (FFA + IOES, Fig. 2.2c, e; Table 2.8). Thus, IOES failed to improve the quality of muscle reinnervation. However, the total number of motor endplates did not differ either between any of the experimental groups or between the experimental groups and intact animals.

2.2.3
Postoperative Electrical Stimulation of Paralyzed Vibrissal Muscles Did Not Improve Recovery of Function

Another potential intervention is the POES of denervated muscles which, maintaining muscle mass and structural integrity, can counteract loss of muscle

Table 2.8 Quality of target muscle reinnervation after facial nerve injury and intraoperative electrical stimulation (IOES)

Group of animals	Mono-innervated motor end-plates (percent)	Poly-innervated motor end-plates (percent)	Non-innervated motor end-plates (percent)	Total number of motor end-plates examined
1. Intact rats	100 ± 0	0	0	1543 ± 132
2. 4 months after FFA+SS	46 ± 9.5[#]	51 ± 10[#]	2.7 ± 1.8[#]	1362 ± 134
3. 4 months after FFA+IOES	38 ± 7.1[#]	42 ± 4.1[#]	20 ± 4.2[#]	1179 ± 240

Innervation pattern of the levator labii superioris (LLS) motor endplates in intact rats (Intact), in rats subjected to transection and suture of the facial nerve (FFA) and 1 h intraoperative sham stimulation (FFA + SS) or electrical stimulation (FFA + IOES) of the proximal stump of the transected facial nerve. At least eight animals were studied per group. Shown are group mean values ± SD

[#]Indicate differences in mean values between intact rats and the experimental groups (ANOVA for repeated measurements and post hoc Tukey's test, $p < 0.05$) and show the expected changes in the quality of motor endplate innervation with no changes in their total numbers

excitability and muscle atrophy resulting from disuse (Kern et al. 2005; Salmons et al. 2005; Ashley et al. 2007, 2008; Salmons and Jarvis 2008). However, evidence has yet to be presented as to whether, and to what degree, preservation of a larger muscle mass and better functional properties of denervated muscles would promote functional recovery after facial nerve reconstruction.

2.2.3.1

Animal Groups and Overview of Experiments

Eighty rats were distributed in five groups each of 16 animals. Group 1 consisted of intact rats and groups 2–5 of experimental rats that were subjected to unilateral transection and suture of the right facial nerve (FFA; Fig. 2.1a). Animals in group 3 (Resection) underwent removal of approximately 1-cm nerve length from the three main branches of the facial nerve (see below). Rats in group 4 (operated, but sham stimulated, FFA + SS) had electrodes inserted in the denervated vibrissal muscles, but no electric current was applied (Fig. 2.6b). In group 5, the vibrissal muscles were subjected to ES (Fig. 2.6c).

Vibrissal motor performance during explorative whisking was analyzed in all rats using VBMA. Following the functional analysis at 2 months after operation, half ($n = 8$) of the animals in all groups were used to establish the degree of collateral axonal branching using triple retrograde neuronal labeling. The remaining rats ($n = 8$) were used to determine the proportion of monoinnervated and polyinnervated motor endplates in the m. levator labii superioris (LLS) using immunocytochemistry for neuronal class III β-tubulin, AChE, and histochemistry for alpha-bungarotoxin (see below).

Transection and suture of the facial nerve (FFA) was performed as described (Sect. 2.1.1).

2.2.3.2
Resection of the Facial Nerve

The main trunk of the facial nerve was unilaterally mobilized at its emergence from the stylomastoid foramen, and 8–10 mm length of the temporal, zygomatic, buccal, upper and lower divisions of the marginal mandibular branch were removed. This resection of the facial nerve is a very severe lesion in comparison with crush or transection of the nerve and delivers a permanent separation of the facial motoneurons from their target musculature.

2.2.3.3
Postoperative Electrical Stimulation

Operated rats were subjected to electrical stimulation (ES) of the vibrissal muscles three times a week (Monday, Wednesday and Friday) over 2 months starting on the first day after surgery. Electrical stimulation was delivered three times weekly since animals were required to be anesthetized. Under ketamin/xylazin anesthesia, two acupuncture needle electrodes were inserted toward levator labii superioris LLS, one along the uppermost vibrissal row A and the other in the lowest row D (Arvidsson 1982). The site of electrode placement close to the nose of the animal was thus somewhat distant (approximately 1 cm) to the location where the majority of LLS motor endplates are found, i.e., at the borderline between the cheek and whisker pad. The configuration thus allowed ES of the target muscles, the LLS, and part of the intrinsic vibrissal muscles close to the stimulating electrode, without the risk of direct damage to the motor endplates.

In all electrically stimulated rats ($n = 16$), the threshold voltage required to elicit visible contractions of the whisker pad muscles and movements of the whiskers was initially determined by applying square 0.1 ms pulses at various voltage intensities using an isolated pulse stimulator (Master-8-cp, A.M.P.I., Jerusalem, Israel). The frequency selected (5 Hz) resembled the frequency of normal whisking. The muscles were stimulated for 5 min by applying square 0.1 ms pulses with supra-threshold amplitudes (typically 3.0–5.0 V; Fig. 2.6c). This stimulation was sufficient to depolarize intramuscular axons but not muscle fibers, innervated or denervated, in which action potentials can be elicited only upon much "stronger" stimulation, e.g., pulses of 20 V amplitude and 5 ms duration for normal muscles and higher for denervated muscle fibers (Irintchev et al. 1990; Kern et al. 2002). Efficient muscle stimulation, especially that of denervated muscle fibers, requires delivery of high-voltage current pulses of long duration. This stimulation protocol was not approved by the Animal Welfare Committee in Cologne because of the concern that "strong" stimulation might elicit trigeminal pain. Control sham-stimulated rats were treated identically to rats subjected to ES except that no current was applied to the electrodes (Fig. 2.6b).

2.2.3.4
Analysis of Vibrissae Motor Performance

Analysis of vibrissae motor performance was performed as already described. Compared with intact animals, vibrissal motion was poor in rats receiving sham stimulation (FFA + SS, Group 4) or postoperative POES (FAA + POES; Group 5) of the vibrissal muscles. The mean amplitude was reduced to 16% and 20% of normal (fourth column in Table 2.9) and the angular velocity to 26% and 17% (fifth column in Table 2.9).

We also noted a specific time course of changes in the intensity of muscle contractions throughout the 2-month period of ES in which the stimulation parameters remained unchanged. Muscle contractions and vibrissal movements were readily visible during the first postoperative week. Thereafter they declined to zero (probably due to anterograde Wallerian axonal degeneration) and appeared again as reinnervation of endplates took place after 2–3 weeks. After the third postoperative week, it became increasingly difficult to elicit muscle contractions similar to those observed at the beginning of the treatment period (data not shown).

2.2.3.5
Application of Three Fluorescence Tracers, Fluorescence Microscopy and Counting

Application of three fluorescence tracers, fluorescence microscopy, and counting were performed as described above. Two months after facial nerve cut/anastomosis and sham stimulation (FFA + SS, Group 4), myotopic organization into subnuclei was no longer observed, i.e., all retrogradely labeled motoneurons were scattered throughout the facial nucleus (Fig. 2.2b). The same phenomenon was observed when FFA was combined with POES (Group 5). The lack of myotopy presumably arose because of poor axonal pathfinding and misdirection of the now

Table 2.9 Motor recovery after facial nerve injury and postoperative electrical stimulation (POES)

Group of animals	Frequency (in Hz)	Angle at maximal protraction (in degrees)	Amplitude (in degrees)	Angular velocity during protraction (in degrees/sec)
1. Intact	7.0 ± 0.8	62 ± 13	57 ± 13	1238 ± 503
2. FFA-only	6.3 ± 0.5	91 ± 12	19 ± 6	135 ± 54
3. Resection	7.0 ± 0.8	102 ± 16	16 ± 5	323 ± 170
4. FFA + SS	5.8 ± 0.7	99 ± 11	16 ± 2.4	323 ± 81
5. FFA + POES	5.6 ± 1.1	106 ± 33	20 ± 8	211 ± 93

Biometrics of vibrissae motor performance in intact rats (Intact), in rats subjected to transection and suture of the facial nerve (FFA-only), in rats that underwent removal of 1 cm length from the main branches of the facial nerve (Resection paradigm), in rats subjected to FFA plus postoperative sham stimulation (FFA + SS) or postoperative electrical stimulation (FFA + POES). All groups consisted of 16 rats. Shown are group mean values ± SD. No significant differences between the control group 4 (FFA + SS) and the group with electrically stimulated rats (ANOVA and post hoc Dunnett's test, $p < 0.05$) were detected. Data for groups 1–3 are given as reference and were not included in the analysis highly branched regenerating axons after transection and suture of the facial nerve. No fascicular orientation in the zygomatic, buccal, or marginal mandibular branches occurred.

Double and triple labeling was also commonly observed (Fig. 2.2a). In addition, robust collateral branching at the lesion site resulted in the retrograde labeling of more motoneuronal perikarya in each of the individual facial nerve branches than in intact animals (Table 2.10). Such increases in axonal numbers, or their branches, in turn lead to hyperinnervation of peripheral muscle targets (Angelov et al. 1996; Streppel et al. 1998).

Retrograde tracing did not reveal any changes in the total numbers of single-labeled motoneurons (i.e., DiI-only + FG-only + FB-only; columns 1, 5, and 6 of Table 2.10) in the facial nucleus. The total number of single-labeled cells was 2,184 ± 242 in intact rats (group 1), 4,281 ± 830 after sham stimulation (FFA + SS, group 4), and 4,731 ± 756 following POES (FFA + POES, group 5).

The index of axonal branching following FFA and either of the treatments (SS or POES) was 57–62% (sum of the percentages of DiI + FG and DiI + FB retrogradely labeled perikarya in the third and fourth column of Table 2.10). In summary, the complete lack of myotopic organization, and a consistently elevated degree of axonal branching, suggested that POES did not influence axonal projection patterns.

Pattern of target muscle reinnervation was estimated as described earlier.

In intact animals, all motor endplates were monoinnervated (Fig. 2.2d, f). After facial nerve injury, the proportion of polyinnervated motor endplates was 49 ± 9.4% after sham stimulation (FFA + SS, Group 4) and 55 ± 14% after POES (FFA + POES, Fig. 2.2c, e; Table 2.11). Thus, POES failed to improve the quality of muscle reinnervation.

A very interesting and novel finding is the massive reduction, compared with other treatments, in the number of motor endplates, identified by the alpha-bungarotoxin-binding AChR, in the LLS at 2 months after ES. The total numbers of motor endplates observed in animals subjected to SS was 1,398 ± 415. Following POES, this number was reduced to approximately 24% of the value in the SS-group.

This observation has several possible explanations. First, many muscle fibers in the electrically stimulated vibrissal muscles have not been innervated, have degenerated, and have disappeared rapidly (Schmalbruch et al. 1991; Dedkov et al. 2001). Alternatively, it is quite possible that the vast majority of muscle fibers in ES muscles persisted, though strongly atrophied, in a denervated state. It has been shown (though in humans) that muscle fibers persist in an atrophic state for several years in denervated muscles before they are replaced by connective tissue (Sunderland 1950; Schwarting et al. 1984). In favor of this possibility was the observation that the diameters of most muscle fibers in FFA + POES rats appeared smaller than in animals subjected to FFA + SS. Finally, it could be that the atrophied fibers seen were actually newly formed, having regenerated in the absence of the nerve and were therefore without endplates. On this basis, POES has promoted regeneration of muscle fibers.

To find more direct evidence for denervated atrophied muscle fibers, we stained representative sections from all electrically stimulated animals with

Table 2.10 Degree of collateral branching after facial nerve injury and postoperative electrical stimulation

Group of animals	Neurons projecting only into the zygomatic nerve (DiI-only)	Neurons projecting into the zygomatic and buccal nerves (DiI + FG)	Neurons projecting into the zygomatic and marginal mandibular nerves (DiI + FB)	All DiI labeled neurons projecting into the zygomatic nerve (DiI, DiI + FG, DiI + FB)	Neurons projecting only into the buccal nerve (FG-only)	Neurons projecting only into the marginal mandibular nerve (FB-only)
1. Intact	364 ± 47	–	–	364 ± 47	1,441 ± 101	379 ± 94
	100%	0%	0%	100%		
2. FFA-only	213 ± 53	239 ± 52	257 ± 56	709 ± 178	1,908 ± 289	1,488 ± 356
	30%	34%	36%	100%		
3. Resection	0	0	0	0	0	0
4. FFA + SS	228 ± 165	159 ± 98	138 ± 79	525 ± 342	2,172 ± 256	1,881 ± 409
	43%	30%	27%	100%		
5. FFA + POES	321 ± 120	237 ± 102	276 ± 83	834 ± 305	2,254 ± 374	2,156 ± 262
	38%	28%	34%	100%		

Number of motoneurons with axons in the zygomatic, buccal, or marginal mandibular branches of the facial nerve in intact rats (Intact), in rats subjected to transection and suture of the facial nerve (FFA-only), in rats that underwent removal of 1 cm length from the main branches of the facial nerve (Resection paradigm), in rats subjected to FFA plus postoperative sham stimulation (FFA + SS) or postoperative electrical stimulation (FFA + POES). The percentage values below the absolute numbers in columns 2–5 indicate the portions of motoneurons projecting through the zygomatic nerve with branched (DiI + FG or DiI + FB, column 3 and 4) and unbranched axons (DiI-only, column 2). Animals were studied 10 days after triple retrograde labeling. At least eight animals were studied per group. Shown are group mean values ± SD. No significant differences between the control group 4 (FFA + SS) and the group with electrically stimulated rats (ANOVA and post hoc Dunnett's test, $p < 0.05$) were detected. Data for groups 1 and 2 are given as reference and were not included in the analysis

Table 2.11 Quality of target muscle reinnervation after facial nerve injury and postoperative electrical stimulation (POES)

Group of animals	Monoinnervated motor endplates (%)	Polyinnervated motor endplates (%)	Noninnervated motor endplates (%)	Total number of motor endplates in LLS muscle
1. Intact rats	100 ± 0	0	0	1,543 ± 132
2. FFA-only	45 ± 9.6	53 ± 10	2.6 ± 1.8	1,326 ± 413
3. Resection	3.1 ± 0.6	1.6 ± 0.4	95 ± 22	416 ± 113
4. FFA + SS	49 ± 7.7	49 ± 9.4	2.4 ± 0.8	1,398 ± 415
5. FFA + POES	4.4 ± 1.8*	5.5 ± 1.4*	91 ± 25*	346 ± 189*

Innervation pattern of the m. levator labii superioris (LLS) motor endplates in intact rats (Intact), in rats subjected to transection and suture of the facial nerve (FFA-only), in rats that underwent removal of 1 cm length from the main branches of the facial nerve (Resection paradigm), in rats subjected to FFA plus postoperative sham stimulation (FFA + SS) or postoperative electrical stimulation (FFA + POES). At least eight animals were studied per group. Shown are group mean values ± SD. Asterisks indicate difference between groups 4 (FFA + SS) and 5 (FFA + POES) at 2 months after surgery (ANOVA and post hoc Dunnett's test, $p < 0.05$). Data for groups 1–3 are given as reference and were not included in the analysis

both alpha-bungarotoxin and goat anti-AchE (Jevsek et al. 2004). Binding of anti-AchE was visualized by anti-goat IgG CY3 conjugate. We expected to see both innervated (alpha-bungarotoxin- and esterase-positive) endplates and denervated endplates outlined by AChE, an enzyme which can persist, in contrast to the AChRs, for months after denervation of muscles limb and trunk muscles (Lomo and Slater 1980; Decker and Berman 1990). Surprisingly, our analysis showed that postsynaptic AChE was present only in association with postsynaptic nicotinic AChRs. The only explanation for this unexpected result was that endplates in facial muscles, in contrast to limb musculature (Gordon et al. 2007, 2008), lose rapidly, within less than 2 months, both endplate markers, AChE and receptors.

To prove this possibility, we permanently denervated all vibrissal muscles in a group of rats ($n = 16$) by resection of the facial nerve. Two months after the operation, the LLS muscles were indeed completely denervated as indicated by the reduction in the total number of motor endplates and by the absence of beta-tubulin-positive axons.

Altogether, our results show that the POES treatment leads to partial muscle reinnervation after FFA, a procedure normally followed by complete muscle innervation (Angelov et al. 1996).

2.2.4

Manual Stimulation of Paralyzed Vibrissal Muscles Following Facial Nerve Injury Promoted Full Recovery of Whisking

Following denervation and before reinnervation, several changes also occur within the muscle including loss of muscle bulk and circulation; connective tissue also shrinks and becomes adherent (fibrosis; Eccles 1944; Sunderland 1950; Bardosi et al. 1987). After several months of complete denervation, muscle membrane properties change, becoming relatively nonresponsive to electrical stimulation (Schwarting et al. 1984; Lieber 1992; Stennert et al. 1994). For patients expected to have nerve regrowth after complete denervation, it is important to minimize fibrosis within the muscle connective tissue so that there will be movable muscle structures after muscle reinnervation to allow reacquisition of the contractile proteins that make muscles work (Lömo and Westgaard 1974; Mokrusch et al. 1990; Nix 1990; McCulloch and Nelson 1995). On the basis of clinically established positive benefits of soft tissue massage, supposed to promote muscle blood flow and to keep in optimum condition while awaiting nerve recovery (Hovind and Nielsen 1974; Beurskens 1990; Frach et al. 1992; Coulson 2005), we tested the effect(s) of manual mechanical stimulation of denervated vibrissal muscles after FFA.

2.2.4.1

Animal Groups and Overview of Experiments

One hundred and eight rats were used with one intact control group and six experimental groups (Table 2.12).

Table 2.12 Experimental design chart depicting animal grouping, treatments and parameters investigated

Group of animals	Video-based motion analysis of vibrissae motor performance	Degree of collateral axonal branching as estimated by triple retrograde labeling	Pattern of reinnervation of motor endplates in m. levator labii superioris
1. Intact animals (16 rats)	16	8	8
2. Animals with right FFA (16 rats)	16	8	8
3. Animals with right FFA + EE (16 rats)	16	8	8
4. Animals with right FFA + right MS (32 rats)	32	8	8
5. Animals with right FFA + EE + right MS (16 rats)	16	8	8
6. Animals with right FFA + left MS (six rats)	6	–	6
7. Animals with right FFA + handling (six rats)	6	–	6

Animal grouping and procedures, e.g., facial–facial anastomosis (FFA), dwelling in enriched environment (EE), mechanical stimulation of the vibrissal muscles (MS). In groups 1–5, the animals that underwent video-based motion analysis were subsequently used for estimation the degree of collateral axonal branching. In groups 6 and 7, the animals that were subjected to video-based motion analysis were thereafter used for establishing the pattern of motor endplates reinnervation

Group 1 consisted of 16 intact rats and group 2 of 16 experimental rats which were subjected to unilateral transection and suture of the right facial nerve (FFA) and left to survive for 2 months. In both groups, all rats were used to determine vibrissal motor performance during explorative whisking using VBMA. Thereafter half of these animals were used to establish the degree of collateral axonal branching ipsilaterally by means of retrograde neuronal labeling. Part of the results obtained for group 1 and 2 has already been published (Guntinas-Lichius et al. 2005). The remaining eight rats in both groups were used to determine the proportion of monoinnervated and polyinnervated motor endplates in the ipsilateral levator labii superioris muscle by means of immunocytochemistry for neuronal class III-tubulin and histochemistry with alpha-bungarotoxin (see below).

In groups 3–7, rats underwent unilateral FFA plus subsequent treatments. The animals of group 3 (16 rats) received environmental stimulation for 2 months in an enriched environment (EE). The animals of group 4 (32 rats) received manual stimulation (MS) of the right whisker pad muscles and the animals of group 5 (16 rats) experienced an enriched environment plus manual stimulation of the right whiskers (EE + MS). Vibrissal motor performance, the degree of axonal branching, and patterns of motor endplate reinnervation were then analyzed.

Groups 6 and 7 consisted of six rats each. Animals in group 6 received FFA on the right side and manual stimulation of the intact contralateral (left) whisker pad muscles. Rats in group 7 received no stimulation of the vibrissal muscles but were handled by the experimentor in exactly the same way as occurred during MS except that MS was not used ("handling").

2.2.4.2 Surgery

Transection and end-to-end suture of the right facial nerve (FFA) were performed as described. Animals were kept in different conditions (see below) for 2 months.

2.2.4.3 Standard Housing/Enriched Environment

After surgery, all animals were allowed to recover in individual cages for 24 h. Thereafter, rats from groups 2, 4, 6, and 7 were placed in *standard* cages (425 mm × 266 mm × 185 mm; polycarbonate; Techniplast, Buguggiate, Italy), each cage with two rats.

All 16 rats in group 3 were placed together in specifically designed cages (three cages size 610 mm × 435 mm × 215 mm connected in a row via polycarbonate tunnels; Techniplast, Buguggiate, Italy) where they experienced group living and an enriched environment consisting of horizontal and inclining platforms and various toys (hanging robes, bridges, tunnels, climbing ladders, and balls). Objects and toys were randomly circulated by removing some and adding others during the course of the experiment (cf. van Praag et al. 2000). All 16 rats of group

5 were treated in an identical way, but they received in addition mechanical stimulation of the vibrissal muscles (see below).

2.2.4.4
Mechanical Stimulation of the Vibrissal Muscles

Mechanical stimulation (both, manual as well as environmental) was initiated 1 day after surgery. Rats were daily subjected to gentle rhythmic forward stroking of the right (groups 4 and 5) or left (group 6) vibrissae and whisker pad muscles (Fig. 2.5d, e) 5 days a week. Rats of groups 5 and 6 were manually stimulated for 5 min a day and rats of group 4 were further distributed into four subgroups (4a, 4b, 4c, and 4d) that were stimulated daily for 1 min, 2 min, 5 min, and 10 min, respectively. The pattern of manual stimulation that we selected mimicked the natural active vibrissal movements during whisking, that is, active protraction and passive retraction (Welker 1964; Wineski 1985). Animals rapidly became accustomed to this procedure within 2–3 days and did not show any signs of stress such as freezing or trying to bite, weight loss, or lack of grooming; rather, animals readily cooperated.

2.2.4.5
Handling of the Animals

All six rats of group 7 were subjected to daily "handling". Starting from the first day after FFA, animals were carefully taken by an investigator out of the cage and held as if they were to receive MS for 5 min (Fig. 2.5f). Thereafter, rats were put back in the cages.

Analysis of vibrissae motor performance was performed as described above.

Recovery of vibrissal motion after facial nerve injury (1) alone, i.e., no mechanical stimulation, (2) after mechanical stimulation for 1 and for 2 min only, (3) after stimulation of the contralateral vibrissal muscles, and (4) after postoperative handling was poor most likely due to inadequate muscle function during protraction and retraction (Berg and Kleinfeld 2003): the amplitude of movements and the angular velocity were reduced to less than 40% and 23% of the values in intact animals, respectively (Table 2.13). Frequency of whisking after nerve repair was similar to that in intact rats (see second column in Table 2.13) which may be due to the robust autonomy and capacity of the whisking pattern generator, represented mostly by neurons projecting to the facial nucleus from the brainstem (Popratiloff et al. 2001; Hattox et al. 2002; Veinante and Deschenes 2003).

Manual stimulation of the ipsilateral whiskers for 5 and 10 min daily had a dramatic effect, resulting in a return of normal whisking as indicated by the amplitude of movement (Fig. 2.7a), as well as by the speed during protraction (Table 2.13). Stimulation by an enriched environment did not result in a return of function although combined stimulation (manual and environmental) did (Table 2.13).

Table 2.13 Motor recovery after facial nerve injury and mechanical stimulation

Group of animals	Frequency (in Hz)	Angle at maximal protraction (in degrees)	Amplitude (in degrees)	Angular velocity during protraction (in degrees/s)
1. Intact	7.0 ± 0.8	62.0 ± 13.2*	57 ± 13*	1,238 ± 503*
2. Right FFA[a]	6.3 ± 0.5	91 ± 12#	19 ± 6#	135 ± 54#
3. Right FFA + EE	6.8 ± 0.9	76 ± 6*	26 ± 5#	490 ± 187#
4a. Right FFA + right MS for 1 min	6.5 ± 0.5	89 ± 6.2	13 ± 4	175 ± 68
4b. Right FFA + right MS for 2 min	6.8 ± 0.9	91 ± 10	14 ± 7	159 ± 127
4c. Right FFA + right MS for 5 min	6.6 ± 0.5	66 ± 15*	51 ± 19*	1,019 ± 408*
4d. Right FFA + right MS for 10 min	6.8 ± 0.8	70 ± 11*	36 ± 18*	781 ± 329*
5. Right FFA + EE + right MS	7.8 ± 2.3	65 ± 16*	55 ± 20*	1,124 ± 358*
6. Right FFA + left MS	6.7 ± 1.0	94 ± 9.5#	20 ± 9.5#	368 ± 118#
7. Right FFA + Handling	6.7 ± 0.8	104 ± 10.1#	18 ± 3.4#	316 ± 71#

Biometrics of vibrissae motor performance in intact rats (*Intact*), in rats after transection and suture of the right facial nerve only (*right FFA-only*), in rats subjected to FFA and postoperative dwelling in enriched environment (*right FFA + EE*), in rats that were subjected to FFA and postoperative mechanical stimulation of the right vibrissal muscles *(right FFA + right MS),* in rats subjected to combined treatment *(right FFA + EE + right MS),* in rats that were subjected to FFA and postoperative mechanical stimulation of the left vibrissal muscles *(right FFA + left MS),* and in rats subjected to FFA and postoperative handling (*right FFA + handling*). Groups 1–3 and 5 consisted of eight animals, group 4 of 32 rats, and groups 6 and 7 of six animals. Shown are group mean values ± SD. Significant differences between group mean values (ANOVA and post hoc Tukey's test, $p < 0.05$): * – from FFA, # – from Intact, FFA + MS and FFA + EE + MS
[a]Values adopted from Guntinas-Lichius et al. (2005)

2.2.4.6
Application of Three Fluorescence Tracers, Fluorescence Microscopy and Counting

The estimated index of axonal branching was 70% (Fig. 2.7b). None of the stimulation paradigms, i.e., manual, environmental, or combined, had any significant influence on the projection patterns (Table 2.14). Thus, there was a complete lack of myotopic organization, increased total numbers of projecting motoneurons, and a consistently elevated degree of axonal branching regardless of whether the animals were subjected to any of the stimulation paradigms or not.

2.2.4.7
Pattern of Target Muscle Reinnervation

In intact animals, all motor endplates were monoinnervated (Fig. 2.2d, f). After facial nerve injury and no stimulation, 53% were polyinnervated, i.e., innervated by two or more axons (Fig. 2.2c, e; Table 2.15). However, manual, but not environmental, stimulation significantly reduced the degree of polyinnervated

Table 2.14 Degree of collateral axonal branching after facial nerve injury and mechanical stimulation

Group of animals	Neurons projecting only into the zygomatic nerve (Dil-only)	Neurons projecting into the zygomatic and buccal nerves (Dil + FG)	Neurons projecting into the zygomatic and marginal mandibular nerves	All DiI labeled neurons projecting into the zygomatic nerve (Dil, Dil + FG, Dil + FG)	Neurons projecting only into the buccal nerve (FG-only)	Neurons projecting only into the marginal mandibular nerve (FB-only)
1. Intact[a]	364 ± 47	–	–	364 ± 47	$1{,}441 \pm 101$	379 ± 94
	100%	0%	0%	100%		
2. FFA[a]	213 ± 53	$239 \pm 52^{\#}$	$257 \pm 56^{\#}$	$709 \pm 178^{\#}$	$1{,}908 \pm 289^{\#}$	$1{,}488 \pm 356^{\#}$
	30%	34%	36%	100%		
3. FFA + EE	208 ± 164	$140 \pm 78^{\#}$	$117 \pm 76^{\#}$	465 ± 234	$2{,}871 \pm 268^{*\#}$	$2{,}484 \pm 409^{*\#}$
	45%	30%	25%	100%		
4. FFA + MS for 5 min daily	276 ± 219	$268 \pm 149^{\#}$	$211 \pm 105^{\#}$	$756 \pm 251^{\#}$	$3{,}162 \pm 342^{*\#}$	$2{,}614 \pm 184^{*\#}$
	36%	35%	29%	100%		
5. FFA + EE + MS	351 ± 178	$286 \pm 137^{\#+}$	$174 \pm 113^{\#}$	$810 \pm 256^{\#}$	$2{,}790 \pm 432^{*\#}$	$1{,}986 \pm 210^{*\#\S}$
	43%	35%	12%	100%		

Number of motoneurons with axons in the zygomatic, buccal, or marginal mandibular branches of the facial nerve of intact rats (*Intact*), in rats after transection and suture of the right facial nerve *(FFA-only)*, in rats subjected to postoperative dwelling in enriched environment (*FFA + EE*), in rats that received postoperatve manual stimulation of the vibrissal hairs *(FFA + MS)* and in rats subjected to combined treatment *(FFA + EE + MS)*. The animals were studied 10 days after triple retrograde labeling performed 56 days post-surgery. At least eight animals were studied per group. Shown are group mean values ± SD. Significant differences between group mean values (ANOVA and post hoc Tukey's test, $p < 0.05$): * – from FFA; # – from Intact; + – from FFA+EE; § – from FFA + EE and FFA + MS. The percentage values below the absolute numbers in columns 2–5 indicate the portions of motoneurons projecting through the zygomatic nerve with branched (DiI + FG or DiI + FB, column 3 and 4) and unbranched axons (DiI-only, column 2)

[a]Values adopted from Guntinas-Lichius et al. (2005)

Table 2.15 Quality of target muscle reinnervation after facial nerve injury and mechanical stimulation

Group of animals	Monoinnervated motor endplates (%)	Polyinnervated motor endplates (%)	Noninnervated motor endplates (%)	Total number of motor endplates examined
1. Intact	100 ± 0	0	0	1,543 ± 132
2. Right FFA	45 ± 9.6	53 ± 10	2.6 ± 1.8	1,326 ± 413
3. Right FFA + EE	50 ± 15	41 ± 15	8.9 ± 5.0*	1,411 ± 441
4 d. Right FFA + right MS (5 m daily)	69 ± 7.9*#	22 ± 5.1*#	9.6 ± 3.9*	1,640 ± 338
5. Right FFA + EE + right MS	66 ± 11*	31 ± 10*	2.7 ± 2.0§	1,345 ± 319
6. Right FFA + left MS	38 ± 7	60± 13	2.0 ± 1.6	1,237 ± 249
7. Right FFA + handling	39 ± 6	57 ± 12	5.0 ± 2.1	1,402 ± 235

Reinnervation pattern of the levator m. labii superioris (LLS) motor endplates in intact rats (*Intact*), in rats after transection and suture of the right facial nerve only (*right FFA-only*), in rats subjected to FFA and postoperative dwelling in enriched environment (right *FFA + EE*), in rats subjected to FFA and postoperative manual mechanical stimulation of the right vibrissal muscles *(right FFA + right MS)*, in rats subjected to combined treatment *(right FFA + EE + right MS)*, in rats subjected to FFA and postoperative mechanical stimulation of the left vibrissal muscles *(right FFA + left MS)*, and in rats subjected to FFA and postoperative handling (*right FFA + handling*). Motor endplates were classified as monoinnervated, polyinnervated, or noninnervated according to the number of beta-tubulin-immunoreactive axons that crossed the boundaries of the endplate. Groups 1–5 consisted of eight animals, groups 6 and 7 of six rats. Shown are group mean values ± SD. Significant differences between group mean values (ANOVA and post hoc Turkey's test, $p < 0.05$): * – from FFA; # – from FFA + EE; § – from FFA + EE and FFA-MS. Values for intact rats are given as reference values and not included in the analysis

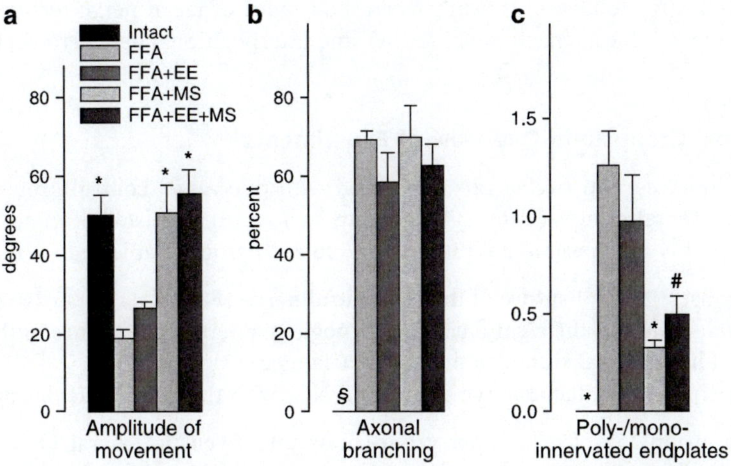

Fig. 2.7 (a–c) Structural correlates of muscle function. Shown are values of major parameters evaluated in (always from *left* to *right*): intact rats (*Intact*), rats subjected to no mechanical stimulation (*FFA*), enriched environment (*FFA + EE*), mechanical stimulation (*FFA + MS*) or combined treatment (*FFA + EE + MS*) for 2 months after facial–facial anastomosis (FFA). Mechanical stimulation (*FFA + MS*), but not enriched environmental housing (*FFA + EE*), leads to full recovery of the amplitude of vibrissal motion (**a**). The extent of axonal branching in the facial nerve trunk is not influenced by any of the stimulation protocols (**b**). The ratio of polyinnervated to monoinnervated motor endplates is strongly reduced as a result of manual but not of environmental stimulation (**c**). The index of axonal branching represents the ratio of motoneurons projecting branched axons into the zygomatic and the buccal or mandibular branch to motoneurons with unbranched axons innervating the zygomatic branch only. Note that the values of the two structural parameters shown are 0 in intact animals. Values are the mean ± SEM. Groups indicated by symbols are significantly different ($p < 0.05$, ANOVA and Tukey's post hoc test) compared with: * – groups FFA and FFA + EE; # – group FFA; § – all other groups. Adopted from Angelov et al. (2007)

endplates (22% and 41%, respectively), whereas the combination of both had an intermediate effect (31%; Table 2.15). Thus, manual stimulation reduced the ratio of polyinnervated to monoinnervated endplates by a factor of 4 compared with untreated rats and to a level, which did not differ statistically from intact animals (Fig. 2.7c).

2.2.5

Manual Stimulation of Facial Muscles Improved Functional Recovery After Hypoglossal–Facial Anastomosis or Interpositional Nerve Grafting

Encouraged by the improvement in function by using MS after FFA (Angelov et al. 2007), we examined whether the same simple rehabilitation technique would be also effective following two other common types of facial nerve reconstruction, hypoglossal–facial anastomosis (HFA), and interpositional nerve grafting (IPNG).

2.2.5.1 Animal Groups and Overview of Experiments

One hundred and twelve rats were used with one intact control group and six experimental groups (Table 2.16). Group 1 ($n = 16$) consisted of intact rats and groups 2–4 of experimental rats ($n = 16$ in each group) subjected to unilateral:

- Transection and suture of the right facial nerve (FFA)
- Transection of the right facial and hypoglossal nerves and suture of the proximal hypoglossal stump to the distal facial nerve stump (HFA)
- Interpositional facial nerve grafting (IPNG) with the great auricular nerve

In groups 5–7 ($n = 16$ per group), rats underwent unilateral FFA, HFA, or IPNG followed by manual mechanical stimulation (MS) of the right whisker pad muscles. Estimation of (1) vibrissal motor performance, (2) degree of axonal branching, and (3) pattern of motor endplate reinnervation was undertaken at 2 months in all experimental groups. Data for stimulated animals were compared with nonstimulated ones.

Vibrissal motor performance during explorative whisking was analyzed in all rats using VBMA. Following functional analysis, half ($n = 8$) of the animals in all groups were used to establish the degree of collateral axonal branching using triple retrograde neuronal labeling. The remaining rats ($n = 8$) were used to determine the proportion of monoinnervated and polyinnervated motor endplates in the ipsilateral levator labii superioris muscle using immunocytochemistry for

Table 2.16 Experimental design chart depicting animal grouping and procedures

Group of animals	Video-based motion analysis of vibrissae motor performance	Degree of collateral axonal branching as estimated by triple retrograde labeling	Pattern of reinnervation of the motor endplates in m. levator labii superioris
1 Intact animals (16 rats)[a]	16	8	8
2. Animals with FFA-only (16 rats)[a]	16	8	8
3. Animals with HFA-only (16 rats)	16	8	8
4. Animals with IPNG-only (16 rats)	16	8	8
5. Animals with FFA + MS (16 rats)[a]	16	8	8
6. Animals with HFA + MS (16 rats)	16	8	8
7. Animals with IPNG + MS (16 rats)	16	8	8

Animal grouping and procedures, e.g., facial–facial anastomosis (FFA), hypoglossal–facial anastomosis (HFA), interpositional nerve grafting (IPNG), manual mechanical stimulation of the vibrissal muscles (MS). All animals underwent video-based motion analysis. Thereafter, one half were used for estimation the degree of collateral axonal branching and the other half for establishing the pattern of the motor endplates reinnervation

[a]Data adopted from Angelov et al. (2007)

neuronal class III β-tubulin and histochemistry with alpha-bungarotoxin (see below). Some data, i.e., those obtained for groups 1, 2 and 5, have been published previously (Angelov et al. 2007).

Transection and end-to-end suture of the right facial nerve (FFA) were performed as described above.

2.2.5.2
Hypoglossal–Facial Anastomosis

The hypoglossal nerve was exposed and transected distally to its union with the upper root of the ansa cervicalis but proximally to its bifurcation into medial and lateral branches. The facial nerve was transected at its emergence from the foramen stylomastoideum but distal to its posterior auricular branch. The proximal stump of the hypoglossal nerve was then microsurgically sutured to the distal stump of the facial nerve (Fig. 2.8a).

2.2.5.3
Interpositional Nerve Grafting

The right facial nerve was exposed and transected 2–3 mm distal to its emergence from the stylomastoid foramen. The great auricular nerve *(n. auricularis magnus)* was exposed and a 5–6 mm length removed. The great auricular nerve graft was inserted between the transected ends of the facial nerve and microsurgically sutured to each end (Fig. 2.8c).

Manual mechanical stimulation (MS) and *analysis of vibrissae motor performance* were performed as described above.

Recovery of vibrissal motion after FFA, HFA, or IPNG alone, i.e., with no subsequent mechanical stimulation of vibrissal muscles, was poor as demonstrated by inadequate muscle function during protraction and retraction. Although the frequency of whisking was similar to that in intact animals, the amplitude of movement was reduced to less than 40% and the angular velocity during protraction to about 15% of the values in intact animals (Table 2.17).

Manual stimulation had a beneficial effect on the whisking amplitude, resulting in either complete restoration in animals subjected to FFA (from $19^\circ \pm 6^\circ$ to $51^\circ \pm 19^\circ$, as reported previously: Angelov et al. 2007) or to its significant increase in animals subjected to HFA (from $21^\circ \pm 4^\circ$ to $30^\circ \pm 5^\circ$) or to IPNG (from $19^\circ \pm 4^\circ$ to $30^\circ \pm 6^\circ$).

Application of three fluorescence tracers, fluorescence microscopy, and counting were performed as already described.

After FFA, myotopic organization of the facial nucleus into subnuclei was lost, i.e., all retrogradely labeled motoneurons were scattered throughout the entire facial nucleus (Fig. 2.2b). Double- and triple-labeled motoneurons were commonly observed due to the formation of collateral axon branches innervating two or three facial nerve rami resulting in a branching index of 70% (Table 2.18). Each individual nerve ramus therefore contained axons, or axonal branches, of

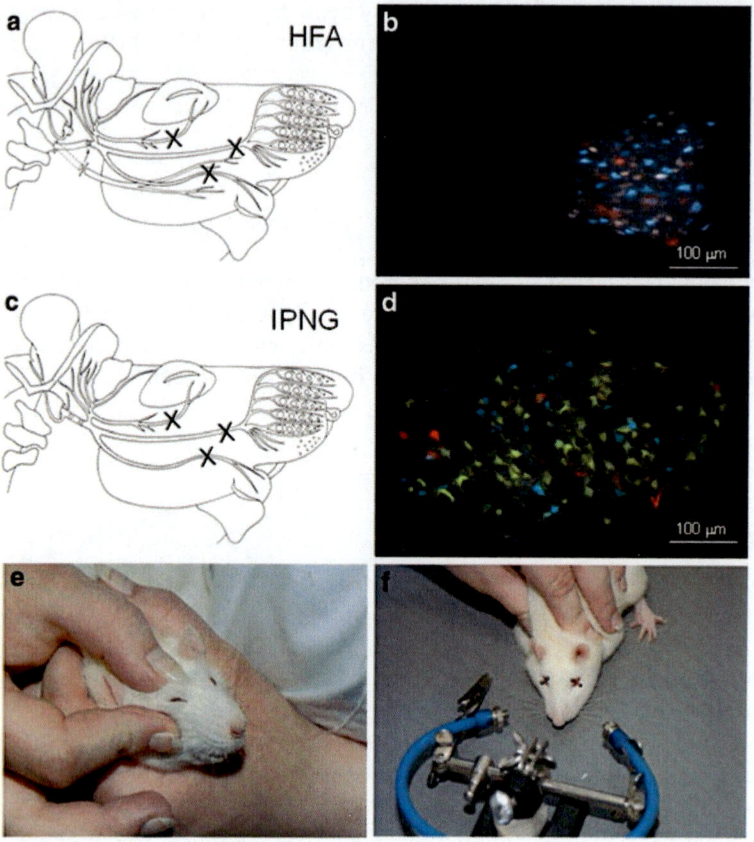

Fig. 2.8 (**a**–**f**) Schematic drawings of the infratemporal portion of the rat facial nerve. Transection of the facial and hypoglossal nerves with subsequent end-to-end suture of the proximal hypoglossal stump to the distal facial fragment indicated by an *arrow* (**a**). Transection of the facial nerve with subsequent end-to-end suture with the interpositional nerve graft (*arrow*) between the proximal and distal facial fragments (**c**). Two months after HFA (**b**) or IPNG (**d**), the myotopic organization is lost irrespective of whether the animals were stimulated or unstimulated. Adopted from Guntinas-Lichius et al. (2007). (**e**) Manual mechanical stimulation of the right, i.e., ipsilateral to the facial nerve transection orbicularis oculi muscle (OOM). (**f**) The blink reflex was evaluated utilizing custom-designed apparatus that delivered a constant 20 ml volume of an air puff to the cornea and periorbital region at a distance of 3 cm. Adopted from Bischoff et al. (2009)

more motoneurons than in intact animals resulting in target hyperinnervation (Rich and Lichtman 1989; Angelov et al. 1996).

After hypoglossal–facial anastomosis and triple retrograde labeling, two major changes, characteristic of aberrant reinnervation of targets, were detected. First, the hypoglossal nucleus lacked somatotopy, i.e., perikarya were not organized into a dorsal subnucleus (with axons projecting into the lateral hypoglossal nerve

Table 2.17 Recovery of whisking function after facial nerve reconstruction and subsequent treatment

Group of animals	Frequency (in Hz)	Angle at maximal protraction (in degrees)	Amplitude (in degrees)	Angular velocity during protraction (in degrees/s)
1. Intact rats[a]	7.0 ± 0.8	62 ± 13	57 ± 13	1,238 ± 503
2. Rats with FFA-only[a]	6.3 ± 0.5	91 ± 12*	19 ± 6.0*	135 ± 54*
3. Rats with HFA-only	6.8 ± 0.6	96 ± 12	21 ± 4.0**	363 ± 177
4. Rats with IPNG-only	7.5 ± 0.8	92 ± 8.2	19 ± 4.0***	406 ± 191
5. Rats with FFA + MS[a]	6.6 ± 0.5	66 ± 15*	51 ± 19*	1,019 ± 408*
6. Rats with HFA + MS	7.6 ± 0.9	91 ± 10	30 ± 5.0**	437 ± 159
7. Rats with IPNG + MS	7.5 ± 0.5	83 ± 16	30 ± 6.0***	382 ± 87

Biometrics of vibrissae motor performance in *intact* animals and in rats that received no postoperative treatment (*FFA-only, HFA-only and IPNG-only*). The other three groups of animals were subjected to daily manual mechanical stimulation (MS) of the vibrissal muscles and hence named FFA + MS, HFA + MS, and IPNG + MS. All groups consisted of 16 animals. Shown are group mean values ± SD. Mean values of a given "stimulated" group (Nr. 5–7) that were significantly different (ANOVA and post hoc Tukey's test, $p < 0.05$) from the respective nonstimulated group (Nr. 2–4) are indicated by *, **, and ***. Values for intact rats are given as reference values and not included in the analysis

[a]Data adopted from Angelov et al. (2007)

branch) and ventral subnucleus (axons projecting into the medial branch of the hypoglossal nerve; Krammer et al. 1979; Uemura-Sumi et al. 1988). Second, the entire hypoglossal nucleus contained double- or triple-labeled motoneuronal perikarya (Fig. 2.8b). Loss of somatotopy was due to transection of the hypoglossal nerve proximally to its bifurcation into its medial and lateral branches and subsequent inaccurate navigation of regrowing neurites into inappropriate branches. Double- and triple-labeled motoneurons arose due to collateral axonal branching at the lesion site. The branching index was 56% (Table 2.18).

Following IPNG and triple retrograde labeling, changes in the facial nucleus were very similar to those observed after FFA: i.e., myotopy was absent and double- and triple-labeled motoneurons were observed throughout (Fig. 2.8c, d). Counts revealed a branching index of 63% (Table 2.18). Following IPNG, the number of retrogradely labeled motoneurons was smaller after applying tracer to the zygomatic (Table 2.18, group 4, fifth column) as compared with the buccal (Table 2.18, group 4, sixth column) and marginal mandibular branches (Table 2.18, group 4, seventh column). The data suggest that IPNG exerts a negative effect on the regrowth of axons into the zygomatic compared with the buccal and marginal mandibular nerves. However, the nature of this possibly mechanical impediment remains unknown.

Mechanical stimulation had no detectable influence on projection patterns after FFA, HFA, and IPNG (Table 2.18). Thus, myotopic organization was completely lacking, the total number of projecting motoneurons was increased (with the exception of ramus zygomaticus after IPNG), and the degree of axonal branching was consistently elevated regardless of whether the animals received MS or not.

Table 2.18 Degree of collateral axonal branching after facial nerve reconstruction and mechanical stimulation

Group of animals	Neurons projecting only into the zygomatic nerve	Neurons projecting into the zygomatic and buccal nerves	Neurons projecting into the zygomatic and marginal mandibular	All DiI labeled neurons projecting into the zygomatic nerve	Neurons projecting only into the buccal nerve	Neurons projecting only into the marginal mandibular nerve
1. Intact[a]	364 ± 47 100%	0 0%	0 0%	364 ± 47 100%	1,441 ± 101	379 ± 94
2. FFA-only[a]	213 ± 53 30%	239 ± 52 34%	257 ± 56 36%	709 ± 178 100%	1,908 ± 289	1,488 ± 356
3. HFA-only	184 ± 125 44%	116 ± 69 28%	116 ± 81 28%	414 ± 211 100%	2,617 ± 623	1,765 ± 1,005
4. IPNG-only	68 ± 28 37%	61 ± 40 33%	56 ± 39 30%	185 ± 100 100%	2,256 ± 376	2,294 ± 519
5. FFA + MS[a]	276 ± 219 36%	268 ± 149 35%	211 ± 105 29%	755 ± 251 100%	1,662 ± 342	1,614 ± 184
6. HFA + MS	231 ± 174 57%	97 ± 40 24%	78 ± 35 19%	406 ± 221 100%	2,224 ± 429	1,743 ± 431
7. IPNG + MS	102 ± 24 48%	62 ± 13 29%	48 ± 21 23%	212 ± 36 100%	2,233 ± 356	2,095 ± 657

Number of motoneurons with axons in the zygomatic, buccal, or marginal mandibular branches of the facial nerve in *intact* animals and in rats that received no postoperative treatment (*FFA-only, HFA-only and IPNG-only*). The other three groups of animals were subjected to daily manual mechanical stimulation of the vibrissal muscles and hence named FFA + MS, HFA + MS, and IPNG + MS. The animals were studied 10 days after triple retrograde labeling performed 56 days post-surgery. The percentage values below the absolute numbers in columns 2–5 indicate the portions of motoneurons projecting through the zygomatic nerve with branched (DiI + FG or DiI + FB, column 3 and 4) and unbranched axons (DiI-only, column 2). All groups consisted of eight animals. Shown are group mean values ± SD. Mean values of a given "stimulated" group (Nr. 5–7) were compared (ANOVA and post hoc Tukey's test, $p < 0.05$) with those of the respective nonstimulated group (Nr. 2–4). No differences were detected. Values for intact rats are given as reference values and not included in the analysis

[a]Data adopted from Angelov et al. (2007)

Table 2.19 Quality of target muscle reinnervation after facial nerve reconstruction and subsequent treatment

Group of animals	Monoinnervated motor endplates (%)	Polyinnervated motor endplates (%)	Noninnervated motor endplates (%)	Total number of motor endplates examined
1. Intact rats[a]	100 ± 0	0	0	1,543 ± 132
2. FFA-only[a]	45 ± 9.6*	53 ± 10*	2.6 ± 1.8	1,326 ± 413
3. HFA-only	68 ± 13**	17 ± 9**	15 ± 6.0	1,524 ± 325
4. IPNG-only	60 ± 11***	22 ± 8***	18 ± 7	1,491 ± 441
5. FFA + MS[a]	66 ± 11*	31 ± 10*	2.7 ± 2.0	1,345 ± 319
6. HFA + MS	79 ± 12**	8 ± 2**	13 ± 5	1,587 ± 402
7. IPNG + MS	76 ± 12***	14 ± 3***	10 ± 3	1,395 ± 312

Reinnervation pattern of the levator labii superioris (LLS) motor endplates in intact rats (*Intact*) and in rats that received no postoperative treatment (*FFA-only, HFA-only and IPNG-only*). The other three groups of animals were subjected to daily manual mechanical stimulation (MS) of the vibrissal muscles and hence named FFA + MS, HFA + MS, and IPNG + MS. Motor endplates were classified as monoinnervated, polyinnervated, or noninnervated according to the number of beta-tubulin-immunoreactive axons that crossed the boundaries of the endplate. All groups consisted of eight animals. Shown are group mean values ± SD. Mean values of a given "stimulated" group (Nr. 5–7) that were significantly different (ANOVA and post hoc Tukey's test, $p < 0.05$) from the respective nonstimulated group (Nr. 2–4) are indicated by *, **, and ***. Values for intact rats are given as reference values and not included in the analysis
[a]Data adopted from Angelov et al. (2007)

Pattern of target muscle reinnervation was estimated as described earlier.

In intact animals, all motor endplates were innervated by one axon and designated as monoinnervated (Fig. 2.2d, f). After FFA, HFA, and IPNG alone (i.e., without stimulation), 53%, 17%, and 22%, respectively, were polyinnervated, i.e., innervated by two or more axons (Fig. 2.2c, e, Table 2.19). Compared with FFA, the decreased degree of polyinnervated motor endplates after HFA and IPNG could be, at least partially, due to more noninnervated endplates after these two types of nerve reconstruction (fourth column of Table 2.19).

Manual stimulation of the vibrissal muscles significantly reduced the proportions of polyinnervated endplates (FFA: 31%, HFA: 8% and IPNG: 14%, respectively; Table 2.19).

2.2.5.4

Manual Stimulation of the Orbicularis Oculi Muscle Improved Eyelid Closure After Facial Nerve Injury in Adult Rats

Following facial nerve injury in humans, soft tissue massage improves blood flow, facial symmetry, and smiling. Together, the findings suggested that interventions that reduced the degree of endplate polyinnervation might also improve functional outcome. We therefore tested the effect of manual stimulation of the vibrissal muscles after facial nerve injury (facial nerve transection and immediate end-to-end anastomosis; FFA) in rat. Minor vibrissal motor performance was first noted at 4–5 weeks after FFA and after two further weeks; recovery was complete with function being indistinguishable from that in intact animals. Encouraged by the efficacy of manual stimulation in improving function of facial muscles surrounding the mouth, we decided to check whether the same simple rehabilitation technique would also prove effective for another facial muscle, the orbicularis oculi (OOM). This muscle is also innervated solely by the facial nerve and controls eyelid closure and blinking, both of which can be severely compromised by facial nerve injury in humans and with significant consequences.

Thirty rats were used. Group 1 ($n = 10$) consisted of intact rats and groups 2–3 of experimental rats ($n = 10$ in each group) subjected to unilateral transection and suture of the right facial nerve (FFA). Rats in group 2 (handling control) did not receive manual stimulation but were handled daily by an investigator in an equivalent fashion to rats receiving manual stimulation. Those in group 3 received daily manual stimulation of the right OOM after FFA (Fig. 2.6e). We examined (1) the quality of eyelid closure and (2) pattern of motor endplate reinnervation at 2 months -FFA in both experimental groups. Data for stimulated animals were compared with nonstimulated ones.

Transection and end-to-end suture of the right facial nerve (FFA) as well as *handling* (handling control) were performed as already described.

2.2.5.5 Manual Stimulation of the Orbicularis Oculi Muscle

Manual stimulation was initiated 1 day after surgery. All rats in group 3 were subjected to gentle rhythmic manual closure of the right eye (by slightly pushing both eyelids together and then letting go) for 5 min per day, 5 days a week for 2 months (Fig. 2.6e). The pattern of manual stimulation mimicked natural blinking or eyelid closure. Animals became accustomed to this procedure within 2–3 days and did not show any signs of stress such as freezing or trying to bite, weight loss or lack of grooming; rather, animals readily complied.

2.2.5.6 Video-Based Motion Analysis of Eye Closure

Two months after FFA and daily handling or manual stimulation, both eyes of each animal were simultaneously videotaped for blink responses. The blink reflex was evaluated utilizing custom-designed apparatus that, when activated, delivered a single standardized portion of 20 ml air as a "puff" to the cornea and periorbital region bilaterally at a distance of 3 cm (Terrell and Terzis 1994; Thanos and Terzis 1995; Fig. 2.6f). Ten puffs were delivered sequentially over 120 s. Using a digital camcorder, animals were videotaped during air-puff-evoked eyelid closures. Evaluation of the blink was performed using a 2D Manual Advanced Video System (Motus 2005). For calibration reasons, a ruler laid 20 cm below the recorder (i.e., at a constant angle of 180° to it) was videotaped before each trial. Video sequences were inspected on a screen and frames selected when the eyelid closure on the intact left side was maximal, i.e., when the distance between the eyelids was smallest. A single reference point half-way along the rim of each eye lid was used to measure the mean distance between both eyelids on the left and right side after ten sequential air puffs. Since in intact animals closure of the eyelids is complete, increase in the inter-eyelid distance indicates functional impairment.

In unoperated rats, the curves (Fig. 2.9a) that show the changes in inter-eyelid distance for the left (blue) and right (red) eye over time displayed a strictly parallel course with nearly full closure after each air puff stimulus. Following FFA, the curve for the operated (right) eyelid remained parallel to the x-axis indicating no blink reflex, a pattern which we observed immediately after FFA and at 1 month thereafter (Fig. 2.9b). Manual stimulation improved blinking function at 2 months as evidenced by a return in synchrony of the two curves (Fig. 2.9c).

In intact animals, the mean minimum distance between the eyelids was similar on the left and right side (0.4 ± 0.2 mm and 0.3 ± 0.1 mm, respectively, Fig. 2.9a).

In both experimental groups, the distance on the unoperated (left) side remained, as expected, unchanged compared with intact animals (group 2, handling control: 1 day, 0.6 ± 0.4 mm; 1 month, 0.3 ± 0.1 mm; 2 months, 0.3 ± 0.1 mm; group 3, manual stimulation: 1 day, 0.3 ± 0.1; 1 month, 0.4 ± 0.2; 2 months, 0.2 ± 0.1).

Eyelid closure was severely impaired 1 day after FFA in both experimental groups as indicated by the large increase, compared with the contralateral side, in the minimum eyelid distance (group 2, control handling: 3.3 ± 1.8 mm; group 3, manual stimulation: 3.4 ± 1.2 mm). Thereafter, the degree of eyelid closure proved to be dependant on whether the animals received manual stimulation or not. In group 2 (control handling), the distance between the two eyelids remained unchanged compared with 1 day throughout the observation period (3.2 ± 1.3 mm and 2.7 ± 0.4 mm at 1 and 2 months, respectively). No improvement was seen after FFA with manual stimulation at 1 month (2.8 ± 1.1 mm). However, blink capacity in the manual stimulation group was dramatically improved at 2 months as indicated by the twofold reduction of the minimum eyelid distance compared with handled controls (1.3 ± 0.5 mm vs. 2.7 ± 0.4 mm; $p < 0.05$).

2.2.5.7

Pattern of Target Muscle Reinnervation

As previously described, the OOM is situated around the entire circumference of the palpebral fissure, extending for at least 5 mm from the conjunctival margin of the upper and lower eyelids, and 4 mm from both the medial and lateral canthi (Gong et al. 2003). Observations of immunostained frontal sections through the OOM revealed that most motor endplates are located in the region close to the lateral margin of the orbita.

In intact animals, all neuromuscular junctions (100%) were innervated by one axon and designated as monoinnervated (May 1986). After FFA without manual stimulation, the proportion of polyinnervated, i.e., innervated by two or more axons, neuromuscular junctions was 42 ± 10% (Table 2.20). Daily manual

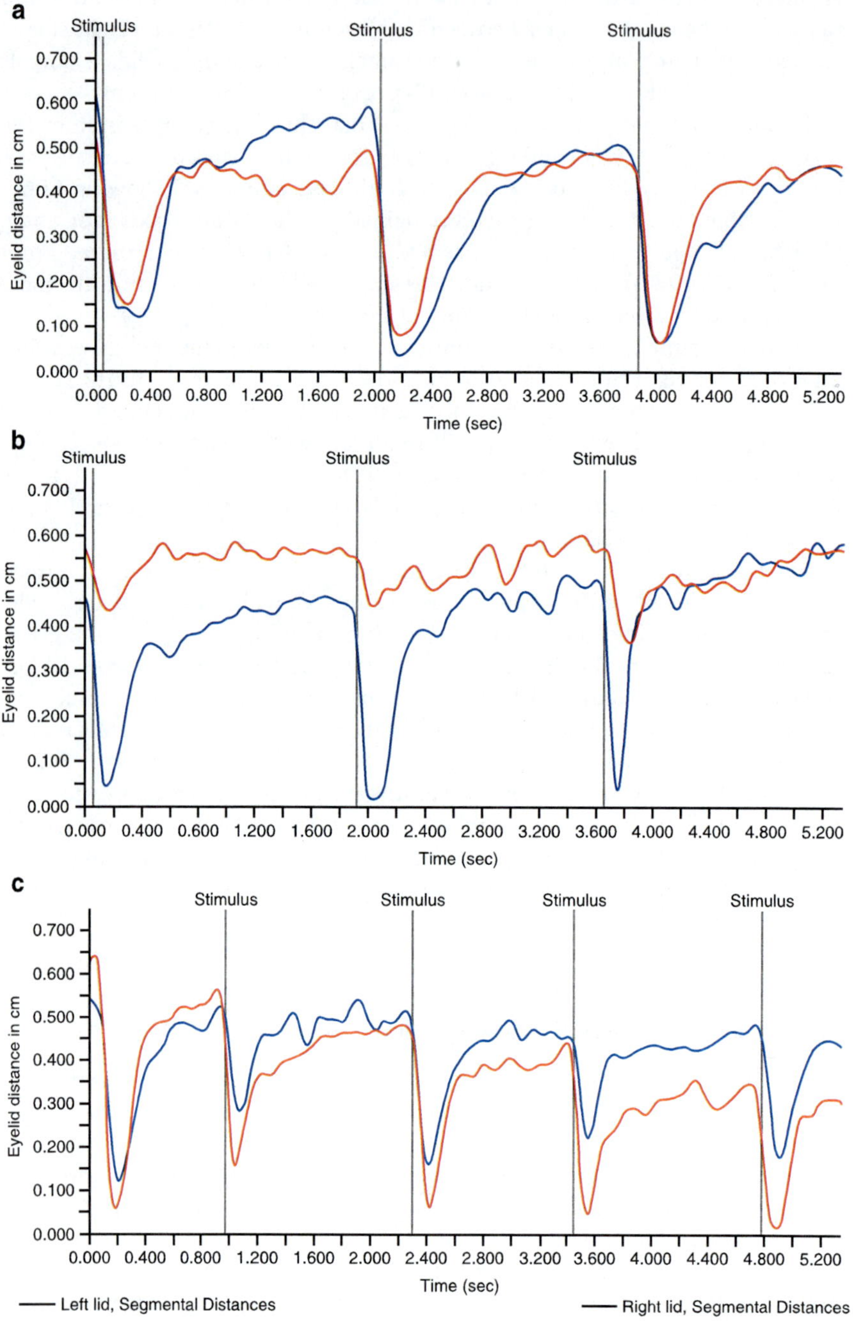

Fig. 2.9 (**a, b**) In unoperated rats, both curves, indicating the closure of the eyelids of the left (in *blue*) and right (in *red*) eye, display a parallel course with a very good closure (minimum

Table 2.20 Quality of orbicularis oculi muscle (OOM) reinnervation after facial nerve reconstruction and subsequent treatment

Group of animals	Monoinnervated motor endplates (%)	Polyinnervated motor endplates (%)	Noninnervated motor endplates (%)	Total number of motor endplates examined
1. Intact rats	100 ± 0	0	0	3,943 ± 532
2. FFA-only	55 ± 8.6	42 ± 10*	3.0 ± 1.8	4,267 ± 780
3. FFA plus manual stimulation	76 ± 9.1*	21 ± 10*	3.0 ± 1.0	3,612 ± 991

Reinnervation pattern of the orbicularis oculi muscle (OOM) motor endplates in intact rats (Intact) and in rats that received no postoperative treatment (FFA-only). The animals of the third group were subjected to daily manual mechanical stimulation of OOM. Motor endplates were classified as monoinnervated, polyinnervated, or noninnervated according to the number of beta-tubulin-immunoreactive axons that crossed the boundaries of the endplate. All groups consisted of ten animals. Shown are group mean values ± SD. Mean values of the "stimulated" group (Nr. 3) that were significantly different (ANOVA and post hoc Tukey's test, $p < 0.05$) from the nonstimulated group (Nr. 2) are indicated by *. Values for intact rats are given as reference values and not included in the analysis

stimulation improved the pattern of reinnervation. Two months after FFA, the fraction of polyinnervated neuromuscular junctions was by a factor of 2 lower than in handled controls (21 ± 10%, $p < 0.05$; Table 2.20).

2.2.6

Manual Stimulation of the Suprahyoid–Sublingual Region Diminished Polyinnervation of the Motor Endplates and Improved Recovery of Function After Hypoglossal Nerve Injury in Rats

Studies in experimental animals have shown that mild electrical stimulation of the denervated soleus muscle inhibits intramuscular sprouting and diminishes motor endplate polyinnervation (Love et al. 2003). In addition, soft tissue massage has been shown clinically to have several benefits (Coulson 2005). The findings prompted us to examine the effect of manual stimulation on both functional recovery of vibrissal muscles and the degree of polyinnervation following facial nerve injury. VBMA of vibrissal motor performance showed that manual stimulation resulted in full recovery of whisking which was associated with reduced polyneuronal reinnervation of motor endplates (Angelov et al. 2007). Since manual

Fig. 2.9 (continued) inter-eyelid distance) after each air puff stimulus. (**b**) This is in sharp contrast with the situation in operated rats in which the orbicularis oculi muscle on the operated side is, as indicated by the lack of blink reflex responses, paretic (red curve) even 2 months after FFA and no manual stimulation. (**c**) Definite improvement of the eye closure on the right side (*red* curve almost parallel to the *blue* one) 2 months after FFA and manual stimulation of the orbicularis oculi muscle. Adopted from Bischoff et al. (2009)

stimulation significantly improved functional recovery of whisking after injury to the facial nerve, we asked whether this rehabilitation approach would be also successful after injury to another nerve, namely the hypoglossal nerve. An additional impetus for this study was the recent finding that motor control of human tongue movements can be improved by selected tongue training techniques (Svensson et al. 2003).

2.2.6.1

Animals, Groups, and Overview of Experiments

Seventy-eight rats were divided into two control and four experimental groups (Table 2.21). Groups 1–3 ($n = 16$ per group) were used to study collateral axonal branching at the site of lesion, synaptic input to the hypoglossal motoneurons, and the quality of target reinnervation (*m. hyoglossus*). Group 1 consisted of intact animals. All rats in groups 2 and 3 were subjected to unilateral transection and suture of the right hypoglossal nerve (hypoglossal–hypoglossal anastomosis, HHA). Animals in group 2 received no postoperative treatment, whereas those in group 3 were subjected to manual stimulation (MS) of the extrinsic and intrinsic suprahyoid–sublingual region.

In addition to misdirected reinnervation of muscle targets, insufficient recovery has also been attributed to rearrangement of cortical representations (Sanes et al. 1988; Svensson et al. 2006). Cortical tongue muscle representation volume was therefore examined (groups 4–6; $n = 10$ per group). Animals in group 4 (intact) were subjected to right unilateral HHA and were kept under anesthesia for

Table 2.21 Experimental design chart depicting animal grouping and procedures

Group of animals	Restoration of the tongue position by measuring the deviation angle	Degree of collateral axonal branching as estimated by double retrograde labeling	Extent of synaptic input to hypoglossal motoneurons	Reinnervation pattern of the motor endplates in the hyoglossus muscle	Tongue muscles representation volume (CTMRV)
1. Intact	16	8	8	8	
2. HHA-only	16	8	8	8	
3. HHA + MS	16	8	8	8	
4. Intact	10				10
5. HHA-only	10				10
6. HHA + MS	10				10

Animal grouping and procedures, e.g., hypoglossal–hypoglossal anastomosis (HHA), with or without manual mechanical stimulation of the tongue muscles (MS). All animals from groups 1–3 were subjected to postoperative measurement of the tongue tip deviation from the midline. Thereafter, one half were used for estimation the degree of collateral axonal branching and the other half for establishing the extent of synaptic input to the hypoglossal motoneurons and the pattern of the motor endplates reinnervation. The animals from groups 4–6 served to establish changes in cortical tongue muscles representation volume (CTMRV)

1 h prior to perfusion fixation. Animals from groups 5 and 6 underwent HHA and survived for 2 months; those in group 5 received no postoperative treatment while those in group 6 received MS exactly as those in group 3. After 2 months, the right hypoglossal nerve of all rats in groups 5 and 6 was transected proximally to the earlier lesion kept under anesthesia for 1 h prior to perfusion fixation.

Analyses of (1) deviation angle of the tongue tip from the midline, (2) degree of axonal branching, (3) synaptic input to the hypoglossal motoneurons, (4) pattern of motor endplate reinnervation, and (5) determination of cortical tongue muscle representation volume were performed at 2 months after surgery. Data for rats receiving MS were compared with those that did not.

All animals in groups 1–6 were used to determine the deviation angle of the tongue tip from the midline, a standard procedure for estimating hypoglossal nerve function (Lowe 1981). Thereafter, half the animals ($n = 8$) in groups 1–3 were used to establish the degree of collateral axonal branching by means of double retrograde neuronal labeling (see below). The remaining rats in groups 1–3 ($n = 8$) were used to determine the synaptic input to the hypoglossal motoneurons (using immunocytochemistry for synaptophysin, see below) and the proportion of monoinnervated and polyinnervated motor endplates in the ipsilateral hyoglossus muscle (using immunocytochemistry for neuronal class III β-tubulin and histochemistry with alpha-bungarotoxin, see below). All animals in groups 4–6 (group 4: intact; group 5 HHA no MS; group 6: HHA + MS) were used to determine the deviation angle of the tongue tip from the midline and the cortical tongue muscle representation volume (see below).

2.2.6.2

Surgery

Transection and end-to-end suture of the right hypoglossal nerve (hypoglossal-hypoglossal anastomosis, HHA) were performed after intraperitoneal injection of ketamin/xylazin. The right hypoglossal nerve was exposed and transected proximal to its bifurcation into lateral and medial branches (Fig. 2.10a). End-to-end suture (hypoglossus–hypoglossus anastomosis, HHA) was performed immediately using two 11–0 atraumatic sutures.

2.2.6.3

Manual Stimulation of the Extrinsic and Intrinsic Suprahyoid–Sublingual Region

On the day following surgery, the suprahyoid–sublingual region of all 16 animals from group 3 and 6 were manually stimulated. MS was performed by gently stroking the lower jaw and upper neck to stimulate all three extrinsic (skeletal) muscles of the tongue *(m. styloglossus, m. genoioglossus, m. hyoglossus)* for 5 min a day, 5 days a week for 2 months. The pattern of manual stimulation mimicked the natural active movements during swallowing (Fig. 2.10b). In addition, upon return to the cage, each rat which had received MS also had a drop of honey

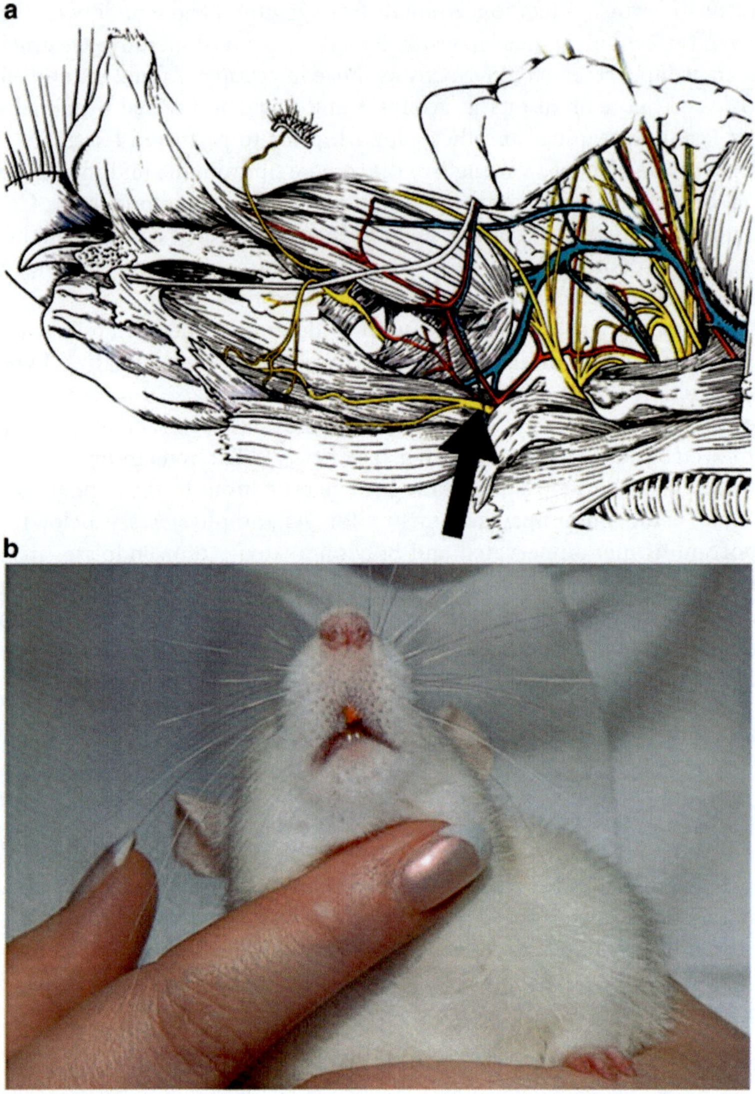

Fig. 2.10 (**a, b**) Surgical procedure and postoperative treatment. (**a**) Schematic drawing of the rat hypoglossal nerve. *Arrow* points at the site of transection and suture. Adopted from Greene (1955). (**b**) Manual "submental" stimulation of the extrinsic suprahyoid–sublingual region. Adopted from Evgenieva et al. (2008)

placed on its back which was accessible to its cage companions. By licking away the honey, cage companions stimulated their intrinsic suprahyoid–sublingual region (*m. longitudinalis sup., m. longitudinalis inf. m. transversus and m. verticalis*) for additional 5–10 min after MS.

2.2.6.4

Restoration of Tongue Position During Protrusion As a Sign for Recovery of Function

Recovery of tongue function was estimated by measuring deviation of the tongue tip from the midline, i.e., the angle between the long axis of the tongue and the median line of the body running between the incisor teeth. Animals were held gently by an experimentator and the upper lip was slightly lifted. Photographs in the frontal plane of all rats were taken from an identical distance (about 20 cm) using the macro-menu of a Nikon 50D digital camera. The very high resolution of the pictures allowed us to readily identify the tip of the tongue, as well as the long axis of the organ. The identical position of each animal when photographed and the short distance between the camera and the head of the rat reduced the possible parallax errors maximally.

In intact rats, tonus of the right and left protruders (the extrinsic *m. genioglossus* and the intrinsic vertical and transverse muscles) is identical, and therefore the tip of the tongue was situated exactly in the midline behind the lower incisors, i.e., the deviation from the midline was 0° (Fig. 2.11a). Following right-sided HHA, malfunction of all right protruders resulted in domination of the opposite (left) muscles which displaced the tongue tip to the right (Wilson et al. 1994), i.e., the long axis of the tongue no longer coincided with the midline (Fig. 2.11b).

Daily MS of the suprahyoid–sublingual region 5 min a day for 2 months improved tongue position after HHA (Fig. 2.11). Deviation of the tongue tip following MS was significantly lower compared with nonstimulated animals ($37.4 \pm 9.37°$ vs. $50.1 \pm 9.01°$; mean ± SD, $n = 26$, Mann–Whitney test, $p = 0.026$).

2.2.6.5

Estimation of Axonal Branching by Double Retrograde Labeling

Previous data (Angelov et al. 1994) after immunostaining of 50-μm-thick vibratome sections for neuron-specific enolase (NSE, i.e., no retrograde labeling performed) showed that the intact hypoglossal nucleus contained 3,576 ± 284 NSE-immunoreactive perikarya; there were no significant changes in these values either at 1 (4,010 ± 245) or 8 weeks after HHA (3,412 ± 348).

2.2.6.6

Application of Fluorescence Tracers

Eight rats from groups 1–3 were used to analyze the degree of collateral axonal branching at the lesion site (HHA). Under Rompun/Ketanest anesthesia, the right hypoglossal nerve was re-exposed distally to the suture site. The medial and lateral branches were transected and instilled with crystals of the retrograde fluorescent dyes Fast Blue and Fluoro-Gold, respectively (Fig. 2.12a). Crystals were left in situ for 30 min after which the application sites were carefully rinsed, dried, and the

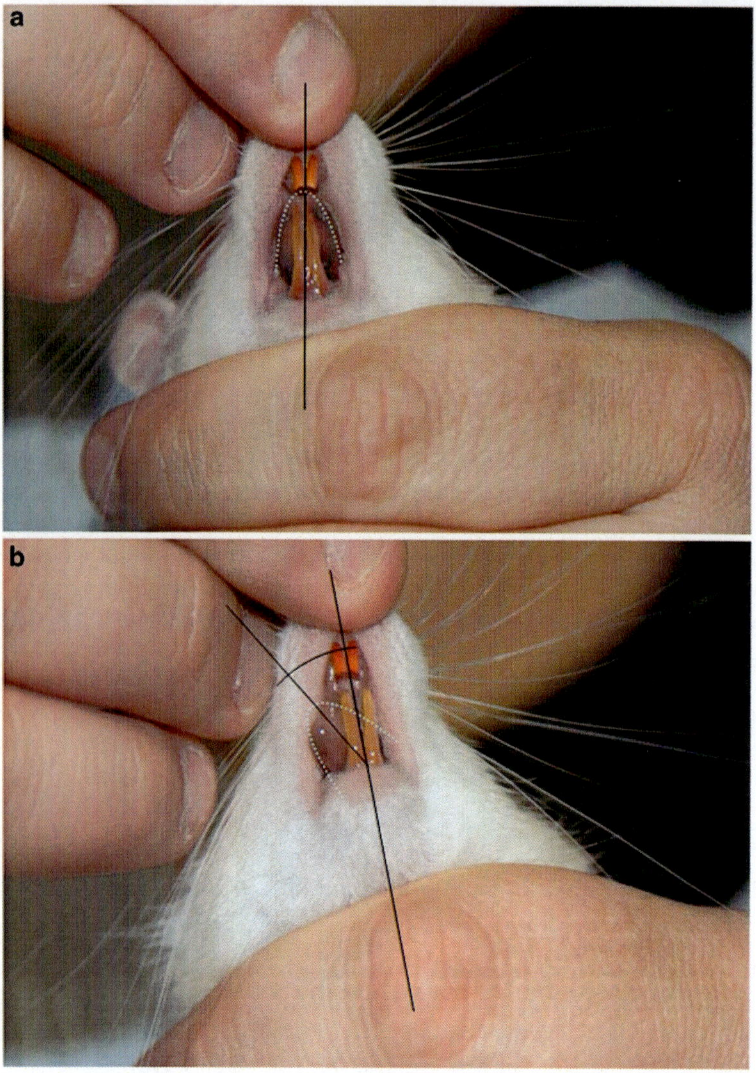

Fig. 2.11 (**a**, **b**) Measurement of tongue tip deviation from the midline, i.e., of the angle between the long axis of the organ and the median line of the body running between the incisor teeth, in representative animals. The edges of the tongue are outlined by a *dotted line*. (**a**) In intact rats, the identical tonus of the right and left protruders and transverse muscles situated the tip of the tongue exactly in the middle behind the lower incisors, i.e., the deviation from the midline was 0°. (**b**) In operated animals, the left protruder dominated and displaced the tongue tip to the right, i.e., the long axis of the organ was no more covering the median line and the angle between them was proportional to the recovery of function. Adopted from Evgenieva et al. (2008)

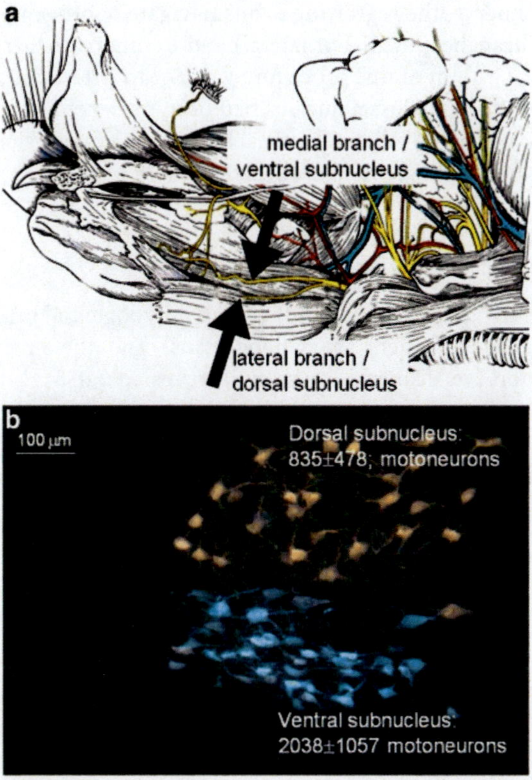

Fig. 2.12 (**a, b**) Retrograde neuronal labeling in intact rats with two fluorescent dyes. (**a**) Schematic drawing of the rat hypoglossal nerve. The *upper arrow* points at the medial branch that was transected and instilled with crystals of FB and the *lower arrow* at the lateral branch of the hypoglossal nerve (transected and labeled with crystals of FG). (**b**) Myotopic organization of the hypoglossal nucleus in intact rats. Application of FB to the transected medial branch labeled perikarya, which were localized in the ventral hypoglossal subnucleus. Likewise application of FG to the transected lateral branch labeled perikarya located in the dorsal hypoglossal subnucleus. No double-labeled perikarya were observed, i.e., the degree of axonal branching was 0%. Adopted from Evgenieva et al. (2008)

wound closed. Ten days later, animals were fixed by perfusion with 4% paraformaldehyde and the brainstems were sectioned coronally at 50 μm.

Under normal physiological conditions, the hypoglossal nerve controls tongue movements by means of its two functionally different nerve branches (Lowe 1981). The medial branch contains the axons of neurons in the ventral hypoglossal subnucleus and innervates muscles that are related to protrusion of the tongue (the extrinsic genioglossus and the intrinsic vertical and transverse muscles). The smaller lateral branch contains the axons of perikarya in the dorsal hypoglossal subnucleus and innervates muscles related to tongue retraction (extrinsic styloglossus and hyoglossus and intrinsic superior and inferior longitudinal). After transection of the hypoglossal nerve, the regrowing axons navigate poorly and fail to rejoin their original nerve branches (medial or lateral) and to innervate therefore their correct muscle targets. The aim of this procedure was not to determine whether myotopic organization of the hypoglossal nucleus had been preserved or restored, but rather to establish the degree of collateral axonal branching at the lesion site by means of double retrograde labeling and neuronal counts (see below).

2.2.6.7 Fluorescence Microscopy and Counts

In intact rats, retrograde labeling of the medial hypoglossal branch, which innervates the tongue protruders (m. genioglossus) and the intrinsic suprahyoid-sublingual region (vertical and transverse), revealed 2,038 $\pm$ 1,057 (mean $\pm$ SD; $n = 8$) FB-labeled perikarya localized within the ventral hypoglossal subnucleus. Labeling of the lateral branch revealed 835 $\pm$ 478 FG-labeled perikarya located in the dorsal hypoglossal subnucleus (Fig. 2.12b). No double-labeled perikarya were observed, i.e., the index of axonal branching was 0%.

After HHA, two major changes, characteristic of aberrant reinnervation of targets, were detected. First, the hypoglossal nucleus lacked somatotopy, i.e., perikarya were not organized into a dorsal subnucleus (with axons projecting into the lateral hypoglossal nerve branch) and ventral subnucleus (axons projecting into the medial branch of the hypoglossal nerve; Krammer et al. 1979; Uemura-Sumi et al. 1988; Aldes 1995). Loss of somatotopy (Fig. 2.13) was due to transection of the hypoglossal nerve proximally to its bifurcation into its medial and lateral branches and subsequent inaccurate navigation of regrowing neurites into inappropriate branches.

Second, the entire hypoglossal nucleus contained double-labeled (FB + FG) motoneuronal perikarya (*arrows* in Fig. 2.13), which arose due to collateral axonal branching at the lesion site. Following MS, there were 1,234 $\pm$ 592 FB_{only}-, 313 $\pm$ 460 FG_{only}-, and 1,302 $\pm$ 715 FB + FG double-labeled perikarya. In non-stimulated rats, there were 600 $\pm$ 364 FB_{only}-, 632 $\pm$ 722 FG_{only}-, and 1,616 $\pm$ 691 FB + FG double-labeled perikarya (mean $\pm$ SD; $n = 8$). The index of collateral axonal branching did not differ between MS and non-MS groups (46% vs. 56%; Mann–Whitney test, $p > 0.05$).

2.2.6.8 Measurement of Motoneuron Soma Sizes

Earlier work has shown that within 1 week after re-injury of chronically axotomized mouse facial motoneurons, their atrophic cell bodies increase in size and expression of growth-related proteins is enhanced (McPhail et al. 2004b). Thus, motoneuron size after a period of recovery from nerve transection and repair followed by a second axotomy is considered to reflect regenerative capacity and thus the functional state of regenerated motoneurons. Similarly, 3 months after femoral nerve transection and repair in mice, the degree of motor recovery

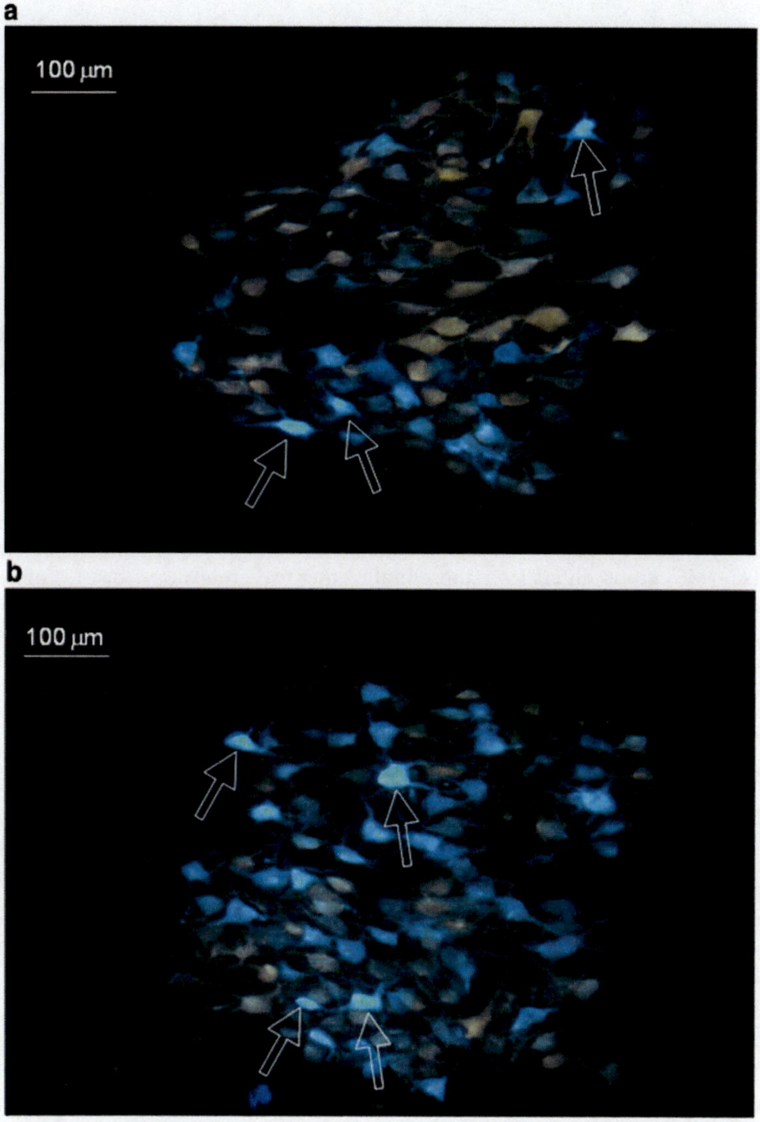

Fig. 2.13 (**a**, **b**) Retrograde neuronal labeling after HHA in nonstimulated (**a**) and stimulated (**b**) rats. The organization of the hypoglossal motoneurons into subnuclei is no longer evident and due to collateral axonal branching there appear double-labeled (FB + FG) neuronal somata (*arrows*). Adopted from Evgenieva et al. (2008)

correlates with soma size of regenerated motoneurons (Simova et al. 2006). We therefore measured hypoglossal motoneuron area following retrograde labeling in rats with and without MS. Labeled hypoglossal motoneurons were photographed with a SPOT-CCD Video Camera System mounted on an Axioplan Zeiss

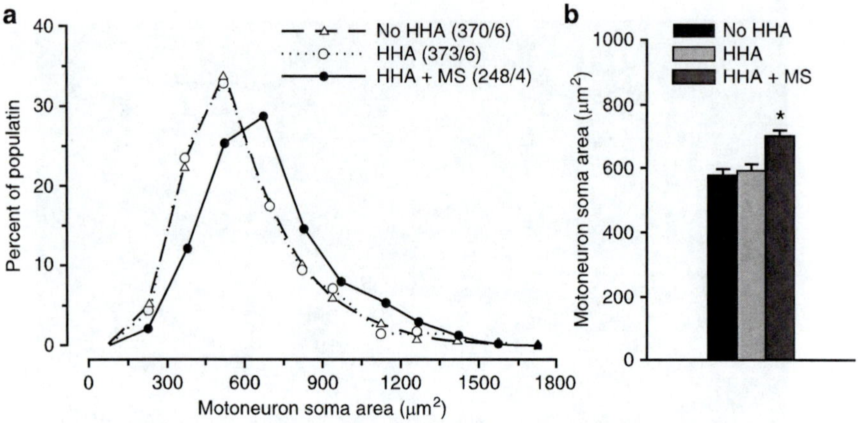

Fig. 2.14 (**a**, **b**) Motoneuron soma size. (**a**) Normalized frequency distributions of soma areas of back-labeled motoneurons in rats subjected to retrograde labeling only ("No HHA") or to retrograde labeling 2 months after nerve repair and no stimulation ("HHA") or nerve repair and manual stimulation ("HHA + MS"). The distribution in the "HHA + MS" group differs from those in the other two groups ($p < 0.001$, Kolmogorov–Smirnov test). Number of motoneurons/rats studied is indicated in the panel. (**b**) Group mean values + SEM of soma areas of the motoneurons shown in (**a**). *Asterisk* indicates significant difference from both other groups ($p < 0.05$, ANOVA with Tukey's post hoc test). Adopted from Evgenieva et al. (2008)

microscope under ×16 magnification. Images were saved in an uncompressed format (TIFF). Analysis was performed with software ImageJ v. 1.38t (US National Institutes of Health, Bethesda, Maryland, USA, http://rsb.info.nih.gov/ij/). An average of 60 motoneurons per hypoglossal nucleus was randomly selected and outlined semiautomatically using the Multi Cell Outliner plugin (Fig. 2.11). The areas of each perikaryon was automatically measured in μm^2 using the building analysis functions of the software. Measurements were performed by one observer (S. Pavlov) who had no information about the postoperative treatment of the rats.

Two months after transection and suture of the right hypoglossal nerve and 10 days after surgery for retrograde labeling, retrogradely labeled motoneuron perikarya were significantly larger in rats receiving MS than in nonstimulated animals ($696 \pm 53\ \mu m^2$ vs. $594 \pm 52\ \mu m^2$, group mean values from individual animal mean values, $p < 0.05$, t test). This conclusion was further verified by analysis of the frequency distributions in the two population samples (Fig. 2.14).

2.2.6.9

Analyses of the Synaptic Input to the Hypoglossal Motoneurons

To compare the degree of synaptic input in rats with and without MS, we measured and established the overall intensity of fluorescence in the hypoglossal nucleus after immunostaining for synaptophysin.

Perfusion fixed (4% paraformaldehyde) brainstems were cut coronally in 30-μm-thick vibratome sections. Immunocytochemical staining for synaptophysin (rabbit polyclonal anti-Synaptophysin, Biometra, Cat. No. 100-599) was performed on every fifth section through the hypoglossal nucleus in one incubation batch for all 24 rats. To quantify pixel brightness, images were captured with a slow scan CCD camera (Spot RT, Diagnostic Instruments Inc., USA; ×16 objective) using Image-Pro Plus Software (Version 5.0; Media Cybernetics, Inc., Silver Spring, MD, USA; Fig. 2.15a–d). Black levels were kept constant but gain was manipulated for each group thereby ensuring that only a few pixels were saturated at the 255 pixel gray value. Each pixel therefore contained 8 bits of information encoding brightness ranging in value from 0 to 255. The scale for pixel brightness, or pixel gray value, was constructed so that the higher numbers indicate greater pixel brightness. The use of the collection filter further reduced the number of pixels saturated at 255. Thus the background intensities were identical from image to image around a pixel gray value of 50. Accordingly, the level for measuring pixel number and brightness was set at 51 (Fig. 2.15e).

To assess changes in total synaptic input to the hypoglossal nucleus in the three groups (fourth column in Table 2.21), we quantified synaptophysin expression in ten equidistant sections at ×16 magnification. Immunocytochemical staining with anti-synaptophysin revealed numerous small immunoreactive puncta within the neuropil of the hypoglossal nucleus (Fig. 2.15a–d).

The intensity of synaptophysin immunofluorescence differed significantly between the groups ($n = 8$ rats in each). Although there were no differences in the pixel distribution curves (data not shown), statistical analysis of total pixel numbers (gray values 51–210) revealed that MS restored synaptophysin levels to those in normal intact animals (Fig. 2.15d). Specifically, compared with intact animals (group 1), synaptophysin levels were restored in animals receiving MS (group 3) but remained lower in those without MS (group 2; $p = 0.022$); likewise, following HHA, synaptophysin levels were significantly higher ($p = 0.011$) in animals receiving MS (group 3: HHA + MS) compared with those that did not (group 2: HHA-only).

2.2.6.10

Analysis of Target Muscle Reinnervation

The ratio between monoinnervated versus polyinnervated motor endplates was determined as described previously. We selected the hyoglossus muscle rather than the m. genioglossus and m. styloglossus. The hyoglossus muscle extends as a thin muscle sheet from the hyoid bone and enters the tongue laterally, between the masseter and stylohyoideus muscles, allowing its easy identification and dissection (Fig. 2.16a, b). Analysis of the target muscle reinnervation was performed as already described. "intact rats" and the group "HHA + MS" according to one-way ANOVA with post hoc Bonferroni test.

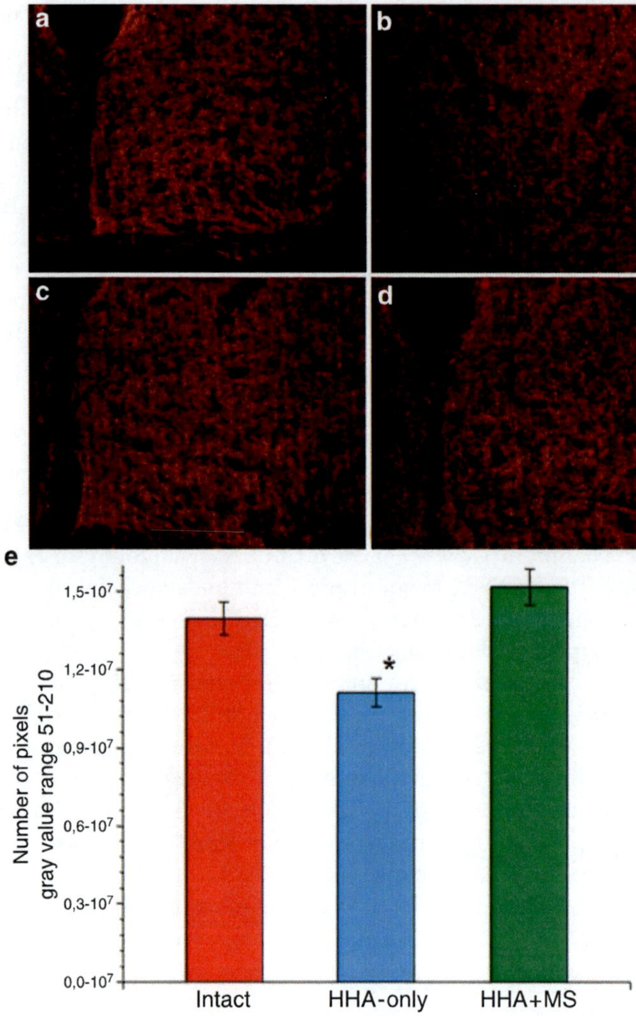

Fig. 2.15 (**a–e**) Quantification of synaptic terminals. Measurements were made using 30-μm-thick vibratome sections through the intact (**a**), contralateral to HHA (**b**), and lesioned hypoglossal nucleus either without (**c**) or with MS (**d**). (**e**) Graphical representations of the intensity of fluorescence in the intact and lesioned hypoglossal nucleus after immunostaining for synaptophysin and Cy3 as florescent dye. Sections were photographed at ×16 magnification. Shown are mean values + SD of pixel numbers within the defined range of *gray* values (51–210). Each experimental group comprised of eight rats. *Asterisk* indicates a significant reduction in the number of pixels in group "HHA-only" when compared with those of group "intact rats" and the group "HHA + MS" according to one-way ANOVA with post hoc Bonferroni test. Adopted from Evgenieva et al. (2008)

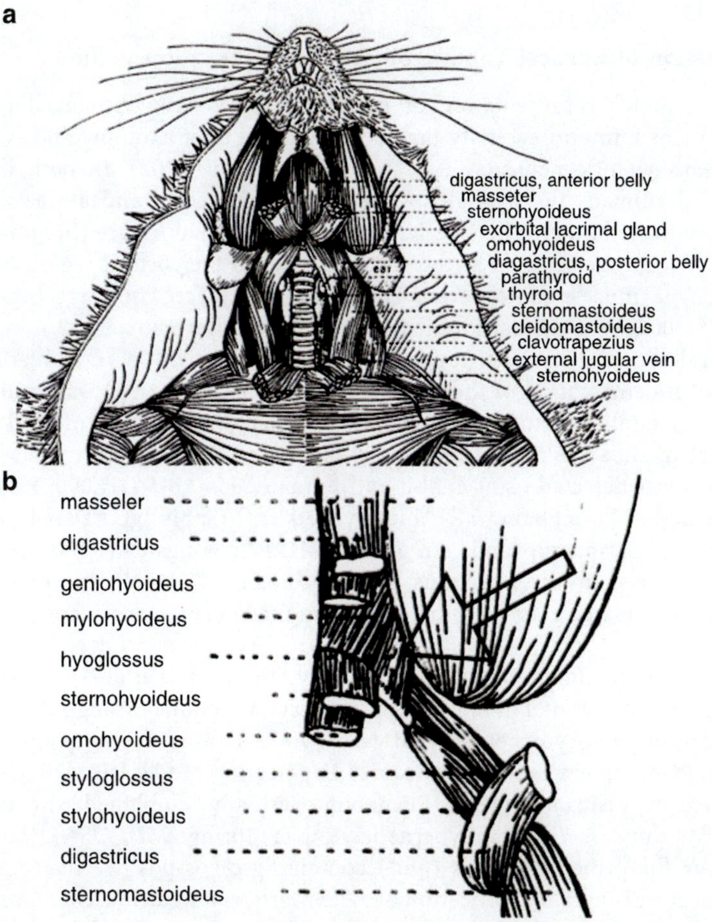

Fig. 2.16 (**a**, **b**) Schematic drawings demonstrating an overview of the supra- and infrahyoid musculature of the rat (**a**) and the detailed localization of the hyoglossus muscle (**b**). While a differentiation among m. geniohyoideus, m. mylohyoideus, m. sternohyoideus, and m. omohyoideus is sometimes hard to achieve, this thin muscle sheet (*arrow*) extends unvariably from the hyoid bone and enters the tongue laterally between the masseter and stylohyoideus muscles. Adopted from Evgenieva et al. 2008

The number of polyinnervated and monoinnervated motor endplates in the hyoglossus muscle of rats receiving MS was, respectively, 327 ± 126 and $1{,}464 \pm 289$; in nonstimulated rats, numbers were 375 ± 151 and 803 ± 329. The number of polyinnervated endplates was significantly smaller in rats receiving MS than in nonstimulated animals (18% vs. 32%; ANOVA and post hoc Bonferroni test, $p < 0.0001$).

2.2.6.11

Estimation of Cortical Tongue Muscle Representation Volume

Cortical motor representation of musculature has been visualized previously using c-Fos immunoreactivity that is upregulated after axotomy and is a marker for *trans*-synaptic neuronal activation (Bisler et al. 2002). In both intact and operated animals, the hypoglossal nerve was transected and animals perfused with fixative 1 h later. The clearly visible ventral rhinal fissure (bregma 5.0 mm; Fig. 2.17a, b) was selected as the rostral border of the cortical region containing the tongue muscle motor area. The brain was placed in a rat brain matrix (RBMS-300C, World Precision Instruments, Berlin, Germany; Fig. 2.17c), allowing identical slices to be cut through the telencephalon with a razor blade in each animal. A demarcation of the side contralateral to HHA (left) was made using a canule perforation in the caudoputamen (Fig. 2.17d). The cryoprotected (sucrose-infiltrated) slice was cut coronally (100 sections; 50 μm thick).

c-Fos was detected using rabbit anti-human cFos-Ab-5 (1:5,000; PC38, Merck Biosciences, Nottingham, UK), biotinylated anti-rabbit IgG (DakoCytomation, Hamburg, Germany), and streptavidin–HRP conjugate (1:100; Amersham, Freiburg, Germany). All sections used to compare CTMRV between stimulated and nonstimulated rats were incubated simultaneously using identical solutions (Fig. 2.17e, f).

Using the fractionator sampling strategy (Gundersen et al. 1988), each tenth coronal section (a total of at least ten equidistant sections through the brain) was used for immunocytochemistry of c-Fos. A Zeiss microscope equipped with a CCD Video Camera System (Optronics Engineering Model DEI-470, Goleta, CA, supplied by Visitron Systems, Puchheim, Germany) combined with Image-Pro Plus 5.0 software (Media Cybernetics, Silver Spring, MD, USA) was used to quantify the projection areas (μm^2) containing c-Fos-positive neurons in each section at a primary magnification of $\times 2.5$. Cortical tongue muscle representation volume was calculated according to the Cavalieri method (Gundersen et al. 1988). Measurements were performed by three observers (P. Schweigert, S.K. Angelova, and D.N. Angelov) who had no information about treatment of the rats.

One hour after transection of the right *intact hypoglossal nerve* (group 4), c-Fos immunopositive nerve cells were seen in the tongue muscle projection area in both the left and right anterior-lateral neocortex (Fig. 2.17), our findings being in agreement with previous studies (Rodel et al. 2004; Donoghue and Wise 1982; Neafsey et al. 1986). Right and left cortical tongue muscle representation volumes did not differ and were 0.063 ± 0.02 and 0.052 ± 0.02 mm^3 ($p < 0.05$).

At 2 months after HHA, we re-transected the right hypoglossal nerve trunk in groups 5 and 6 inducing acute, within an hour, *trans*-synaptic upregulation of c-Fos expression in the motor cerebral cortex (Bisler et al. 2002; Narita et al. 2003; Peeva et al. 2006). As for intact animals, left and right cortical tongue muscle representation volumes did not differ in either MS (left: 0.07 ± 0.03 mm^3; right: 0.08 ± 0.04 mm^3; $p > 0.05$) or nonstimulated (left: 0.12 ± 0.06 mm^3; right:

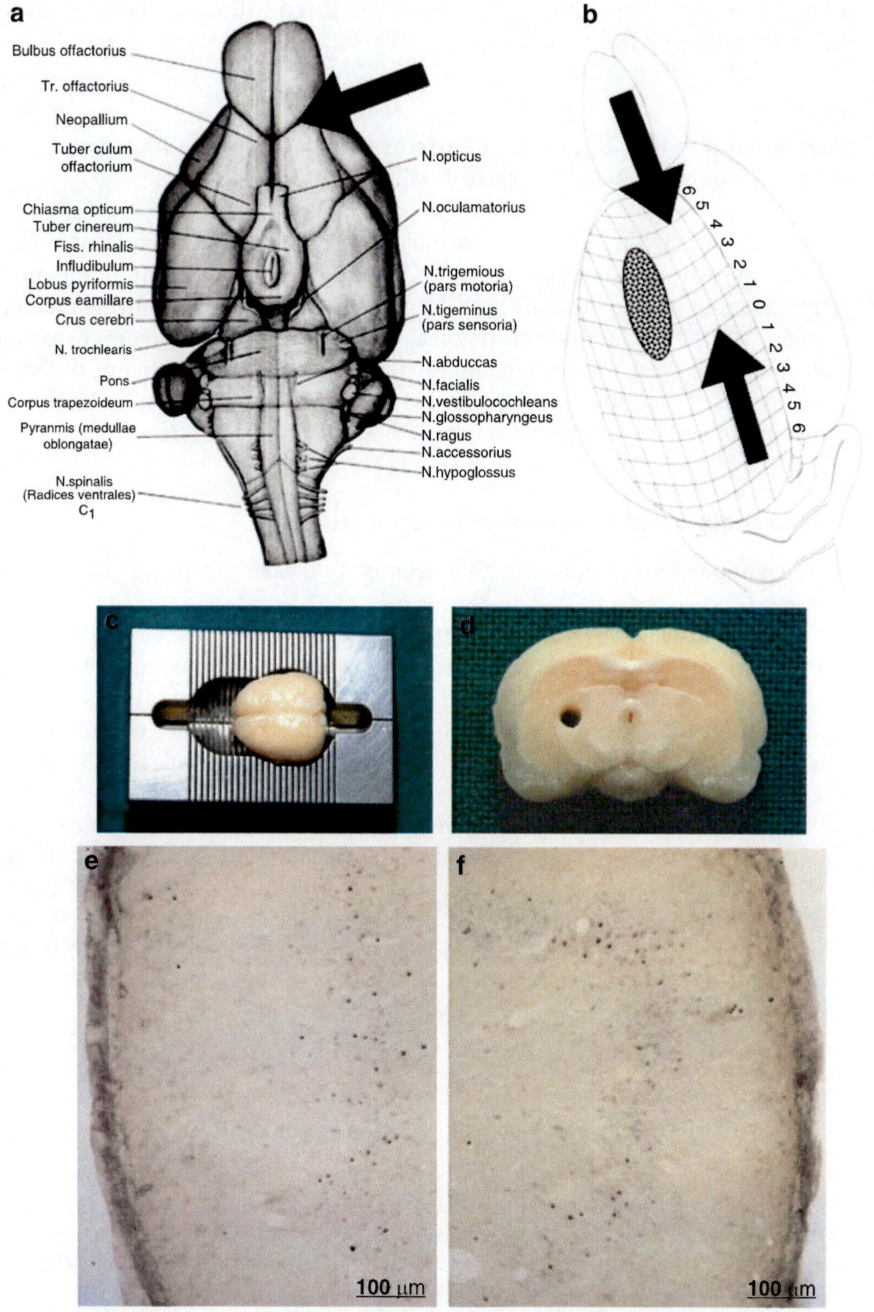

Fig. 2.17 (**a–f**) Quantification of cortical tongue muscle representation volume. (**a**) Ventral aspect of the rat cerebrum with depicted rostral borderline of the brain slice, i.e., the rhinal

0.12 ± 0.05 mm^3; $p > 0.05$). Furthermore, compared with intact rats, values did not differ following MS or in the nonstimulated animals.

2.2.7 Manual Stimulation of Forearm Muscles Did Not Improve Recovery of Motor Function After Injury to a Mixed Peripheral Nerve

It has been recently shown that brief manual stimulation of the whisker pad muscles restored normal whisking function by re-instating monoinnervation of the motor endplates rather than by reducing sprouting at the injury site or restoring myotopy (Angelov et al. 2007; Guntinas-Lichius et al. 2007). Given the effectiveness of manual stimulation following transection and re-anastomosis of a purely motor nerve, we asked to what extent this treatment would influence the outcome after injury to a mixed motor and sensory nerve, namely the median nerve in rats.

2.2.7.1 Animal Groups and Overview of Experiments

Forty-eight rats were divided into four groups (1–4) each consisting of 12 animals (Table 2.22).

Group 1 consisted of 12 intact animals. All rats in groups 2–4 were subjected to unilateral transection and suture of the right median nerve (medianus–medianus anastomosis (MMA)). Animals in group 2 received no postoperative treatment. Animals in group 3 were subjected to manual stimulation (MS) of the forearm whereby the skin and muscles were gently stroked for 5 min a day for 3 months. The protocol was aimed at manually stimulating the muscles within the antebrachium which contains the flexors of the fingers and which are responsible for grip strength (see below III. Surgery). We thus limited MS to the forearm and did not stimulate the skin on the palmar side of the paw which receives primarily sensory innervation from the median nerve (Greene 1955). To control for handling effects, animals in group 4 were only held in hand by the experimentator for 5 min in each session (handling paradigm).

Fig. 2.17 (continued) fissure (bregma 5.0 mm). (**b**) Schematic drawing of the rat brain indicating the dimensions of the slice containing the tongue motor area. The entire brain was placed in a rat brain matrix (RBMS-300C, World Precision Instruments) (**c**) which allowed cutting of identical slices through the telencephalon of all animals. A demarcation of the side contralateral to HHA (*left*) was made using a canule perforation in the caudoputamen (**d**). The cortical representation of the suprahyoid–sublingual region (intact or reinnervated) is in the anterior-lateral neocortex TM1 of both cerebral hemispheres (**e, f**) as identified by c-Fos immunoreactive neurons (with nuclear localization of the DAB-HRP immunoreaction product) 1 h after transection of the right hypoglossal nerve. The portions containing reactive cortical motoneurons were delineated, their areas calculated, and the volume determined according to the Cavalieri principle. Adopted from Evgenieva et al. (2008)

Table 2.22 Experimental design chart depicting animal grouping and procedures

Group of animals	Restoration of grasping force	Degree of collateral axonal branching as estimated by double retrograde labeling	Pattern of reinnervation of the motor endplates in the flexor digitorum profundus muscle
1. Intact animals	12	6	6
2. Rats with MMA-only	12	6	6
3. Rats with MMA + MS	12	6	6
4. Rats with MMA + handling	12	6	6

Animal grouping and procedures, e.g., medianus–medianus anastomosis (MMA), with or without manual mechanical stimulation of the forearm muscles (MS). All animals were subjected to measurement of the grasping force. Thereafter, one half were used for estimation the degree of collateral axonal branching and the other half for establishing the pattern of the motor endplates reinnervation

All animals were used to determine grip force, a standard procedure for estimating motor function of forearm and hand flexors (Bertelli and Mira 1995). Thereafter, half the animals ($n = 6$) were used to establish the degree of collateral axonal branching by means of double retrograde neuronal labeling (see below). The remaining six rats in each group were used to determine the proportion of monoinnervated and polyinnervated motor endplates in the forearm *(m. flexor digitorum sublimis)* using immunocytochemistry for neuronal class III β-tubulin and histochemistry with alpha-bungarotoxin (see below).

2.2.7.2

Anatomy

The rat median nerve runs together with the axillary artery and the radial (musculospiral) nerve to the brachium. At this point, it does not branch, but in the hollow of the elbow two branches arise, one medial and one lateral. The medial branch supplies the *m. flexor digitorum sublimis* and the *m. palmaris longus*; the lateral branch innervates the *m. flexor digitorum profundus* and the *m. pronator quadratus*. Continuing through the antebrachium, the medial branch divides just above the transverse carpal ligament into three common volar digital nerves to the first, second, and third interdigital spaces. Each of these divides into two proper volar digital nerves to the adjacent sides of the first and second, second and third, and third and fourth digits (Greene 1935).

2.2.7.3

Surgery

The right median nerve was exposed in the brachium and transected proximal to its bifurcation into lateral and medial branches (Fig. 2.18a). End-to-end suture (MMA) was performed immediately.

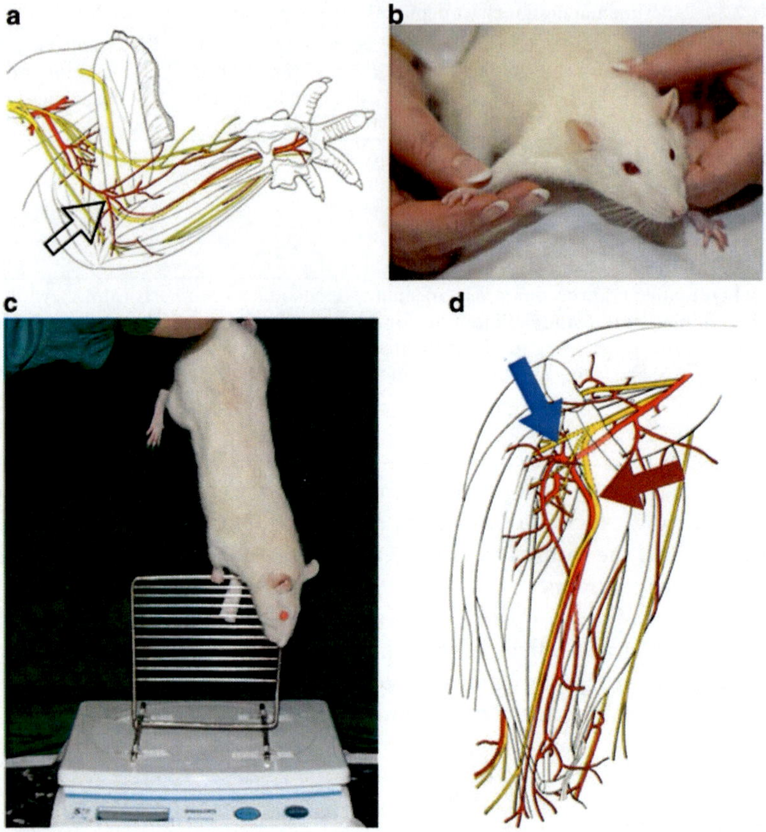

Fig. 2.18 (**a**) Schematic drawing of the rat median nerve. *Arrow* points at the site of transection and suture. (**b**) Manual mechanical stimulation of the forearm flexor muscles. (**c**) Measuring the grip force by means of a wire grid taped onto an electronic balance. (**d**) Schematic drawing of the rat median nerve to depict the retrograde neuronal labeling with two fluorescent dyes. The *blue arrow* points at the medial branch that was transected and instilled with crystals of FB and the *red arrow* at the lateral branch of the median nerve (transected and labeled with crystals of FG). Adopted from Sinis et al. (2008)

2.2.7.4

Manual Stimulation of the Forearm Muscles

On the day following surgery, the right forearms of all 12 animals from group 3 were manually stimulated by stroking all anterior forearm muscles (m. palmaris longus, m. flexor digitorum sublimis, m. flexor digitorum profundus, m. pronator quadratus) that lie just under the skin and the very thin fascia antebrachii. This massage was performed for 5 min daily in a very gentle way, a procedure which avoided postoperative rupture of the coaptation (Fig. 2.18b).

2.2.7.5

Restoration of Grip Force as a Sign for Recovery of Function

The grip test was used to assess functional regeneration. The test is based on denervation of the flexor muscle group which results in a loss of finger flexion after injury to the median nerve. After successful regeneration, the ability to flex the digits and grasp (gripping ability) is regained (Bertelli and Mira 1995; Bontioti et al. 2003; Papalia et al. 2003; Blanco et al. 2007). Grip force was measured using a wire grid (8 × 14 cm) taped to an electronic balance. Normally, rats grasp for the grid when held over it by the tail. When the rats are gently lifted by the tail with increasing strength, they will lose their grip when the maximum grip force is exceeded. The corresponding value displayed on the balance reflects the maximum grip force (Fig. 2.18c). Animals were first given the grip test when active finger flexion was observed by the examiner. Each assessment comprised three grasping attempts and the highest value was recorded. The test was performed weekly for 3 months. To avoid grasping via the left (intact) paw, fingers were covered by a piece of adhesive textile tape. Examiners (D. Bösel and D. Felder) were blinded as to treatment group.

Daily observations of the animals revealed no deviations from their normal behavior, gait, or walking patterns. Despite paresis of the right forepaw finger flexors, rats were able to stand and feed normally.

Daily MS of the forearm muscles 5 min a day for 3 months did not improve recovery of grip function after MMA. Throughout the entire postoperative period, the grip force following MS was not significantly higher compared with nonstimulated animals or to rats subjected only to "handling" (Table 2.23; mean ± SD, $n = 12$, Mann–Whitney test).

2.2.7.6

Estimation of Axonal Branching by Double Retrograde Labeling

Six rats from each group were used to determine the degree of collateral axonal branching at the lesion site (MMA). Under Rompun/Ketanest anesthesia, the right median nerve was re-exposed distally to the suture site. The medial and lateral branches were transected and instilled with crystals of the retrograde fluorescent dyes Fast Blue and DiI, respectively (Fig. 2.18d). Crystals were left in situ for 30 min after which the application sites were carefully rinsed, dried, and the wound closed. Ten days later, animals were fixed by perfusion with 4% paraformaldehyde. The spinal cord and the dorsal root ganglia (DRG) C_5 – Th_1 were cut into 50-μm-thick longitudinal sections.

Sections were observed with an epifluorescence microscope (Zeiss Axioskop 50, Oberkochen, Germany) using a custom-made band pass-filter set combination which maximally limits fluorescence crosstalk between the tracers. Separate color images of retrogradely labeled motoneurons and dorsal root ganglion (DRG) cells were visualized through the different filter sets using a CCD

Table 2.23 Time course of restoration of the grasping force in grams

Group of animals	2 weeks after MMA	4 weeks after MMA	6 weeks after MMA	8 weeks after MMA	12 weeks after MMA
1. Intact animals	264 ± 25	289 ± 33	280 ± 37	293 ± 34	320 ± 43
2. Rats with MMA-only	31 ± 16	60 ± 30	100 ± 36	180 ± 63	218 ± 67
3. Rats with MMA + MS	33 ± 11	48 ± 27	80 ± 45	164 ± 27	232 ± 39
4. Rats with MMA + handling	23 ± 9	50 ± 23	105 ± 49	204 ± 50	239 ± 35

All groups consisted of 12 animals. Shown are group mean values ± SD. Mean values of a given "stimulated" group (Nr. 3, 4) were compared (ANOVA and post hoc Tukey's test, $p < 0.05$) with those of the nonstimulated group (Nr. 2). No differences were detected. Values for intact rats are given as reference values and not included in the analysis

Video Camera System described above. All cells stained by FB_{only}, DiI_{only}, as well as all double-labeled (FB + DiI) cells (Fig. 2.20) were identified and manually counted on the computer screen (Dohm et al. 2000). We quantified the degree (index) of axonal branching (sum of the percentages given in the third and fourth column in Tables 2.24 and 2.25). In rats with an intact median nerve trunk that had been subjected only to surgery for tracer application, the index of axonal branching was 0% (Fig. 2.19).

Counts of all perikarya labeled with FB, DiI, and FB + DiI (Figs. 2.19 and 2.20) were undertaken using the fractionator principle (Gundersen 1986) examining every third section through the spinal cord and spinal ganglion. Details have been described previously (Neiss et al. 1992; see also Valero-Cabre et al. 2004). Counting was performed blindly with respect to treatment.

In intact rats, application of crystals of the fluorescent tracer DiI to the lateral branch of the median nerve (supplying the *flexor digitorum profundus* and *pronator quadratus*) retrogradely labeled 889 ± 61 DRG cells (Fig. 2.19a) and 665 ± 42 motoneurons (mean ± SD; $n = 6$) in the ventral horn of the spinal cord (Fig. 2.19d; Tables 2.24 and 2.25).

Likewise, application of the retrograde fluorescent dye Fast Blue to the medial branch (supplying *flexor digitorum sublimis* and *palmaris longus*) retrogradely labeled 542 ± 93 DRG cells (Fig. 2.19b) and 397 ± 55 motoneurons (Fig. 2.19e; Tables 2.24 and 2.25). No double-labeled neuronal somata were observed either in the DRG (Fig. 2.19c) or in the ventral horn of the spinal cord (Fig. 2.19f). Thus, the index of collateral axonal branching for both sensory and motor neurons was 0%.

Neither MS nor handling changed the degree of postoperative collateral axonal branching. After MMA, double-labeled (DiI + FB) neurons were seen, regardless of the postoperative treatments (Fig. 2.20c, f). In nonstimulated rats (group 2), there were 174 ± 57 (12%) double-labeled DRG cells and 143 ± 21 (16%) motoneurons. A similar index of collateral axonal branching (12–20%) was observed following MS (group 3) or handling (group 4) (Tables 2.24 and 2.25).

Table 2.24 Peripheral projection pattern of sensory (DRG) neurons after median nerve reconstruction and subsequent treatment

Group of animals	Neurons projecting only through the lateral branch (DiI-only)	Neurons projecting only through the medial branch (FB-only)	Neurons projecting through the lateral and medial branches (DiI + FB)	All labeled neurons projecting through the median nerve (DiI, FB, DiI + FB)
1. Intact animals	889 ± 61	542 ± 93	0 0%	1,431 ± 102 100%
2. Rats with MMA-only	613 ± 180	642 ± 121	174 ± 57 12%	1,428 ± 288 100%
3. Rats with MMA + MS	649 ± 229	606 ± 246	254 ± 115 17%	1,509 ± 452 100%
4. Rats with MMA + handling	844 ± 158	591 ± 184	262 ± 57 15%	1,696 ± 217 100%

Number of pseudounipolar neurons (dorsal root ganglion cells) with peripheral processes in the lateral or medial branches of the median nerve in *intact* animals and in rats that received no postoperative treatment (*MMA only*). The animals from the third group were subjected to daily manual mechanical stimulation of the forearm muscles (*MMA + MS*) and those from the fourth group received daily "handling" (*MMA + handling*). The animals were studied 10 days after double retrograde labeling performed 3 months post-surgery. The percentage values below the absolute numbers in column 4 indicate the portions of neurons (DiI + FB) projecting through the median nerve with branched peripheral processes. All groups consisted of six animals. Shown are group mean values ± SD. Mean values of a given "stimulated" group (Nr. 3, 4) were compared (ANOVA and post hoc Tukey's test, $p < 0.05$) to those of the nonstimulated group (Nr. 2). No differences were detected. Values for intact rats are given as reference values and not included in the analysis

2.2.7.7

Measurement of Motoneuron Soma Sizes

Measurement of motoneuron soma sizes was performed as already described. Three months after transection and suture of the right median nerve and 10 days after surgery for retrograde labeling, back-labeled motoneuron perikarya in manually stimulated rats were not significantly larger than in nonstimulated animals subjected to handling only (576 ± 46 μm^2 vs. 586 ± 34 μm^2, group mean values from individual animal mean values, $p > 0.05$, t test). This conclusion was further verified by analysis of the frequency distributions in the two population samples (Fig. 2.21).

2.2.7.8

Analysis of Target Muscle Reinnervation

The ratio between monoinnervated versus polyinnervated motor endplates was calculated as described previously. We selected the superficial head of the *m.*

Table 2.25 Projection pattern of motoneurons after median nerve reconstruction and subsequent treatment

Group of animals	Neurons projecting only through the lateral branch (DiI-only)	Neurons projecting only through the medial branch (FB-only)	Neurons projecting through the lateral and medial branches (DiI + FB)	All labeled neurons projecting through the median nerve (DiI, FB, DiI + FB)
1. Intact animals	665 ± 42	397 ± 55	0 0%	1,062 ± 58 100%
2. Rats with MMA-only	328 ± 68	396 ± 131	143 ± 21 16%	866 ± 165 100%
3. Rats with MMA + MS	302 ± 104	434 ± 105	185 ± 34 20%	921 ± 163 100%
4. Rats with MMA + handling	302 ± 111	409 ± 69	170 ± 56 19%	881 ± 199 100%

Number of motoneuronal perikarya projecting through the lateral or medial branches of the median nerve in *intact* animals and in rats that received no postoperative treatment (*MMA-only*). The animals from the third group were subjected to daily manual mechanical stimulation of the forearm muscles (*MMA + MS*) and those from the fourth group received daily "handling" (*MMA + handling*). The animals were studied 10 days after double retrograde labeling performed 3 months post-surgery. The percentage values below the absolute numbers in column 4 indicate the portions of neurons (DiI + FB) projecting through the median nerve with branched axons. All groups consisted of six animals. Shown are group mean values ± SD. Mean values of a given "stimulated" group (Nr. 3, 4) were compared (ANOVA and post hoc Tukey's test, $p < 0.05$) with those of the nonstimulated group (Nr. 2). No differences were detected. Values for intact rats are given as reference values and not included in the analysis

flexor digitorum sublimis (FDS) muscle because of its easy identification and reliable dissection.

The overall number of motor endplates in the *flexor digitorum profundus* muscle did not differ between the groups (Table 2.26). The proportion of poly-innervated endplates (Fig. 2.2c, e) also did not differ between groups being 10 ± 2% in nonstimulated rats (group 2), 14 ± 4% following MS (group 3) (MS), and 13 ± 8% following handling (group 4).

2.2.8

Manually Stimulated Recovery of Motor Function After Facial Nerve Repair Requires Intact Sensory Input

Both clinical and experimental data show that recovery of function is better following damage of a purely motor nerve compared with mixed peripheral nerves possessing both motor and sensory axons such as the median nerve (Mackinnon et al. 1985; Terzis and Papakonstantinou 2000; Bontioti et al. 2005; Sinis et al. 2005; Kelly et al. 2007). Nerve supply to facial muscles has the distinct advantage

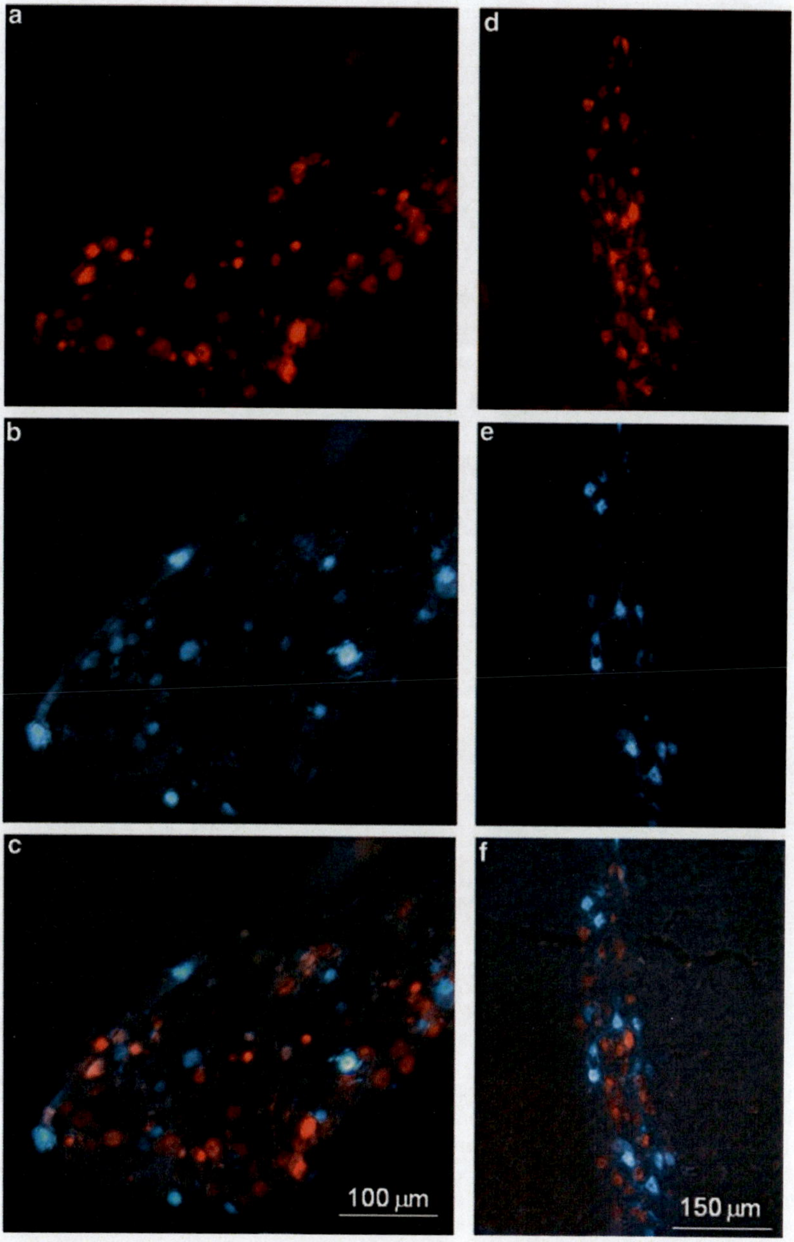

Fig. 2.19 (a–f) Representative cryosection from an intact rat. Application of DiI and FB to the transected lateral and medial branches labeled perikarya, which were localized in the DRG (**a**, **b**) and in the ventral column of the spinal cord (**d**, **f**). No double-labeled perikarya were observed (**c**, **F**), i.e., the degree of axonal branching was 0%. Adopted from Sinis et al. (2008)

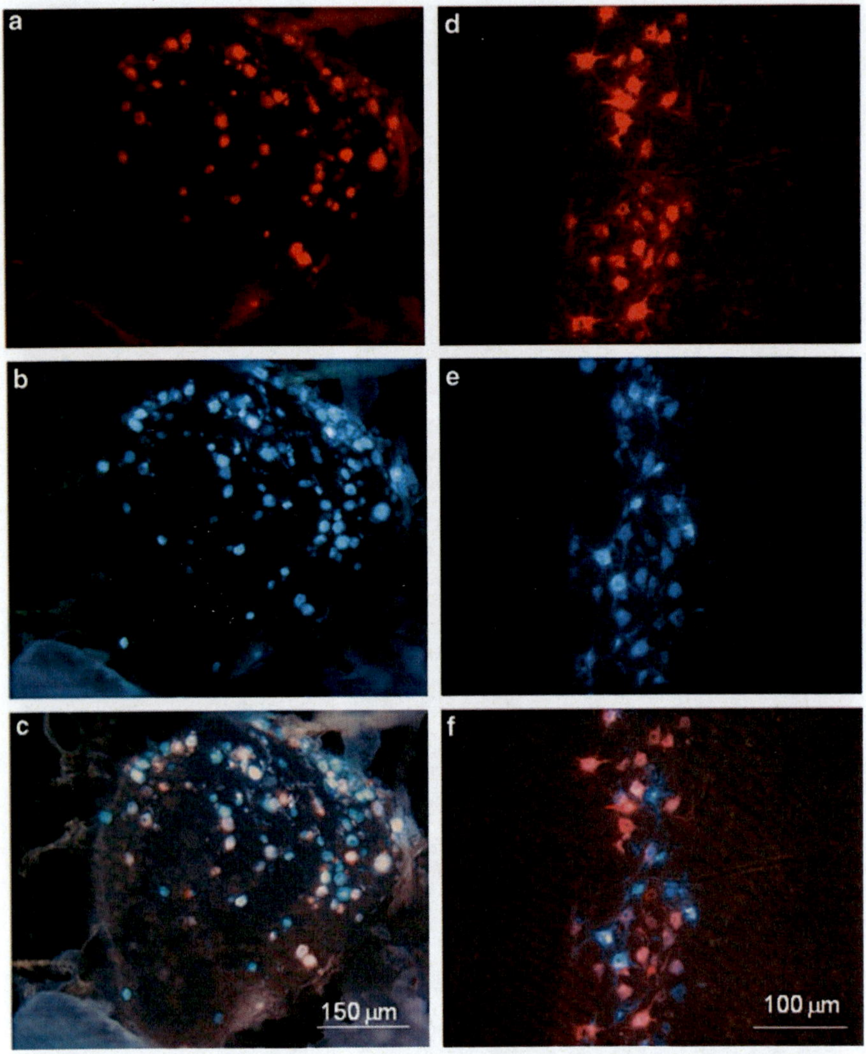

Fig. 2.20 (**a**, **f**) Representative cryosection from an operated rat. Application of DiI and FB to the transected lateral and medial branches labeled perikarya, which were localized in the DRG (**a**, **b**) and in the ventral column of the spinal cord (**d**, **f**). Numerous double-labeled perikarya were observed (**c**, **f**), i.e., the degree of axonal branching was 12–20%. Adopted from Sinis et al. (2008)

that the motor and sensory supplies are separate (Moller and Jannetta 1986; Valls-Sole and Tolosa 1989). In the case of the facial nerve, sensory feed back occurs via the trigeminal nerve with direct ipsilateral connections between the trigeminal and the facial nucleus in the brainstem (Kimura and Lyon 1972; Erzurumlu and

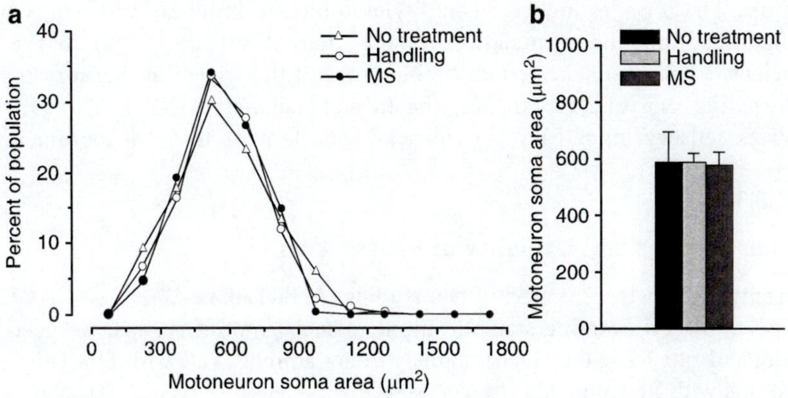

Fig. 2.21 (**a**, **b**) Motoneuron soma size. (**a**) Normalized frequency distributions of soma areas of back-labeled motoneurons in rats subjected to retrograde labeling only ("No treatment") or to retrograde labeling 3 months after nerve repair and handling only ("Handling") or nerve repair and manual stimulation ("MS"). The distribution in the "MS" group does not differ from those in the other two groups ($p < 0.001$, Kolmogorov–Smirnov test). (**b**) Group mean values + SEM of soma areas of the motoneurons shown in **a**. No significant differences between MS and the other two groups ($p < 0.05$, ANOVA with Tukey's post hoc test). Adopted from Sinis et al. (2008)

Table 2.26 Quality of target muscle reinnervation after median nerve reconstruction and subsequent treatment

Group of animals	Monoinnervated motor endplates (%)	Polyinnervated motor endplates (%)	Noninnervated motor endplates (%)	Total number of motor endplates examined
1. Intact animals	100 ± 0	0	0	5,433 ± 1,032
2. Rats with MMA-only	80 ± 10	10 ± 2	5 ± 2	5,962 ± 1,326
3. Rats with MMA + MS	79 ± 13	14 ± 4	7 ± 3	4,865 ± 1,088
4. Rats with MMA + handling	81 ± 11	13 ± 8	6 ± 1	5,362 ± 635

Reinnervation pattern of the flexor digitorum profundus (FDP) motor endplates in intact rats (*Intact*) and in rats that received no postoperative treatment (*MMA-only*). The animals from the third group were subjected to daily manual mechanical stimulation (MS) of the forearm muscles (*MMA + MS*) and those from the fourth group received daily "handling" (*MMA + handling*). Motor endplates were classified as monoinnervated, polyinnervated, or noninnervated according to the number of beta-tubulin-immunoreactive axons that crossed the boundaries of the endplate. All groups consisted of six animals. Shown are group mean values ± SD. Mean values of a given "stimulated" group (Nr. 3, 4) were compared (ANOVA and post hoc Tukey's test, $p < 0.05$) with those of the nonstimulated group (Nr. 2). No differences were detected. Values for intact rats are given as reference values and not included in the analysis

Killackey 1979; Stennert and Limberg 1982; Hinrichsen and Watson 1983; Travers and Norgren 1983; Isokawa-Akesson and Komisaruk 1987; Sharp et al. 1988). We took a two-step approach to examine the role of afferent inputs. First, we estimated, using synaptophysin immunohistochemistry, the influence of manual stimulation on the afferent synaptic input to the facial nucleus. In addition, we tested the influence of the trigeminal sensory input by extirpating one of its branches, the infraorbital nerve (ION). The procedure ablates sensory input from the vibrissal muscle pads to facial motoneurons.

2.2.8.1

Animal Groups and Overview of Experiments

Seventy-eight rats were used in two studies (A, B; Table 2.27)

Study A: To examine synaptic input to facial motorneurons, we used three groups of rats ($n = 6$ in each), namely intact animals, rats with FFA (FFA only), and rats with FFA plus MS (FFA + MS).

Study B: To examine the role of trigeminal afferents (ION) which provide exclusive sensory innervation to the vibrissal muscles (Jacquin et al. 1993; Munger and Renehan 1989; Rice et al. 1993), we used five groups namely intact rats, those with FFA (FFA only), FFA plus MS (FFA + MS), FFA plus excision of the ipsilateral infraorbital nerve (IONex) (FAA + IONex), and those with FFA plus IONex but followed by MS (FFA + IONex + MS). Vibrissal motor performance and the pattern of motor endplate reinnervation were studied at 2 months. Data for animals receiving MS were compared with those lacking MS. Data for intact animals, those with FFA and those with FFA + MS, have been published previously (Angelov et al. 2007).

Table 2.27 Experimental design chart depicting animal grouping, procedures, and investigation mode

Surgery	Motion analysis of vibrissae whisking	Pattern of NMJ reinnervation	Synaptic input to facial motorneurons
Study A			
1. Intact			6
2. FFA only			6
3. FFA + MS			6
Study B			
1. Intact rats[a]	16	8	
2. FFA-only[a]	16	8	
3. FFA + MS[a]	16	8	
4. FFA + IONex	6	6	
5. FFA + IONex + MS	6	6	

Animal grouping, procedures, and investigation mode, e.g., facial–facial anastomosis (FFA), excision of the ipsilateral infraorbital nerve (IONex), manual stimulation of the vibrissal muscles (MS), video-based motion analysis (VBMA), pattern of reinnervation of the neuro-muscular junctions (NMJ), estimation of the synaptic covering of the facial motoneurons after quantitative immunocytochemistry for synaptophysin

[a]Data adopted from Angelov et al., Neurobiol Disease (2007)

2.2.8.2

Surgery

Transection and end-to-end suture of the right facial nerve (FFA) were performed as described. Excision of the ipsilateral ION was performed after FFA. While under ketamin/xylazin narcosis, the ipsilateral ION was transected at its exit from the infraorbital foramen (Fig. 2.4a, b) and all its peripheral fascicles were removed.

2.2.8.3

Manual Stimulation (MS) of Vibrissal Muscles and Analysis of Vibrissae Motor Performance

Manual stimulation (MS) of vibrissal muscles and analysis of vibrissae motor performance were performed as described. Following *FFA alone*, recovery was poor. Although whisking frequency was similar compared with intact animals, both amplitude and angular velocity during protraction were reduced, respectively, to approximately 40% and 15% of values in intact animals (Table 2.28).

However, *MS after FFA* restored the amplitude of vibrissal whisking from 19° ± 6° to 51° ± 19°, that is, within the range for intact animals (Angelov et al. 2007).

By contrast, MS failed to restore vibrissal function in animals which had facial nerve injury, as well as elimination of sensory input *(FFA±IONex±MS)*. Indeed, MS led to a worsening of function with a reduction of whisking amplitude in animals that received MS (FFA + IONex + MS: 14° ± 5.5°) compared with those that did not (FFA + IONex: 22° ± 3.4°; Table 2.28).

Table 2.28 Motor recovery after facial nerve reconstruction and subsequent treatment

Group of animals	Frequency (in Hz)	Angle at maximal protraction (in degrees)	Amplitude (in degrees)	Angular velocity during protraction (in degrees/s)
1. Intact rats[a]	7.0 ± 0.8	62 ± 13	57 ± 13	1,238 ± 503
2. FFA-only[a]	6.3 ± 0.5	91 ± 12*	19 ± 6.0*	135 ± 54*
3. FFA + MS[a]	6.6 ± 0.5	66 ± 15*	51 ± 19*	1,019 ± 408*
4. FFA + IONex	6.0 ± 0.8	76 ± 10	22 ± 3.4	469 ± 400
5. FFA + IONex + MS	6.0 ± 1.2	87 ± 18	14 ± 5.5**	148 ± 68**

Biometrics of vibrissae motor performance in intact animals, in rats that received no postoperative treatment (FFA-only) and in animals that were subjected to FFA plus excision of the ipsilateral infraorbital nerve (FFA + IONex). Groups FFA + MS and FFA + IONex + MS received daily manual stimulation (MS) of the vibrissal muscles. The first three groups consisted of 16 animals and the last two groups of six rats. Shown are group mean values ± SD. Mean values of a given "stimulated" group that were significantly different (ANOVA and post hoc Tukey's test, $p < 0.05$) from the respective nonstimulated group are indicated by * and **. Values for intact rats are given as reference values and not included in the analysis

[a]Data adopted from Angelov et al. (2007)

2.2.8.4

Determining the Synaptic Input to the Facial Motoneurons

Determining the synaptic input to the facial motoneurons was performed as already described. To assess the synaptic input to the facial nucleus in intact rats and rats subjected to FFA with or without subsequent MS (Table 2.27), we

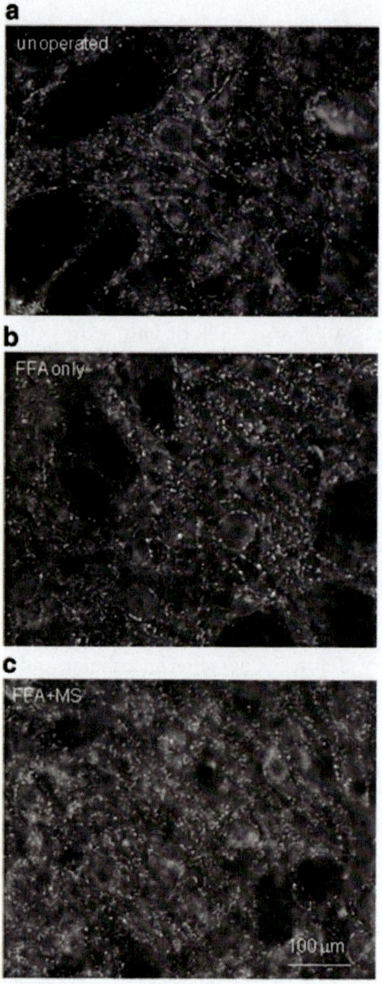

Fig. 2.22 (a–c) Immunostaining for synaptophysin in 30-µm-thick vibratome sections from the facial nucleus in intact rats (**a**), in rats 2 months after facial–facial anastomosis and no further treatment (FFA-only; **b**), and in rats which received MS of the vibrissal muscles after FFA (FFA + MS; **c**). Note the clearly discernible numerous puncta within the neuropil and around motoneuronal cell bodies in the facial nucleus representing synaptic terminals. Adopted from Pavlov et al. (2008)

quantified levels of synaptophysin according to Calhoun et al. (1996) and Marques et al. (2006). Images were obtained on an epi-fluorescence microscope from sections stained with a highly diluted (1:4,000) anti-synaptophysin antibody. This protocol allowed us to obtain photo images in which, comparable with thin confocal optical sections, numerous puncta within the neuropil and around motoneuronal cell bodies in the facial nucleus were clearly discernible (Fig. 2.22).

Analyses of the frequency distributions of pixel intensities revealed no differences among the three groups (data not shown) indicating similar overall intensities of the immunofluorescence staining. The background intensities were identical from image to image around a pixel gray value of 30. Therefore, we used the total number of pixels per frame within the range (gray values 30–129) as an estimate of axon terminal density.

In animals with FFA and no MS, the mean total number of pixels was significantly lower compared with intact animals ($29.2 \times 10^6 \pm 1.8 \times 10^6$ vs. $34.3 \times 10^6 \pm 2.3 \times 10^6$; $p = 0.036$, ANOVA with Bonferroni post hoc test). In animals receiving MS after FFA, the mean pixel number was similar to intact animals ($33 \times 10^6 \pm 2.6 \times 10^6$) but higher than in the control FFA group ($p = 0.007$). These results indicate that the synaptic input after FFA alone is reduced compared with normal and this loss is counteracted by MS.

2.2.8.5
Quality of Target Muscle Reinnervation

Quality of target muscle reinnervation was evaluated in m. levator labii superioris, an extrinsic vibrissal muscle which, like the intrinsic vibrissal muscles, is

Table 2.29 Quality of target muscle reinnervation after facial nerve injury and trigeminal depletion

Group of animals	Monoinnervated motor endplates (percent)	Polyinnervated motor endplates (percent)	Noninnervated motor endplates (percent)	Total number of motor endplates examined
1. Intact rats#	100 ± 0	0	0	1543 ± 132
2. FFA only#	45 ± 9.6	53 ± 10	2.6 ± 1.8	1326 ± 413
3. FFA + MS#	69 ± 7.9*	22 ± 5.1*	9.6 ± 3.9*	1640 ± 338
4. FFA + IONex	51 ± 8.6	43.3 ± 9.4	5.7 ± 2.8	1495 ± 435
5. FFA + IONex + MS	41 ± 6.1	50.7 ± 10	8.3 ± 3.6	1579 ± 443

Innervation pattern of the m. levator labii superioris (LLS) motor-endplates in intact animals, in rats that received no postoperative treatment (FFA-only) and in animals that were subjected to FFA plus excisiion of the ipsilateral infraorbital nerve (FFA + IONex). Groups FFA + MS and FFA + IONex + MS received daily manual stimulation (MS) of the vibrissal muscles. The first three groups consisted of eight animals and the last two groups of six rats. Shown are group mean values ± SD. Mean values of a given "stimulated" group that were significantly different (ANOVA and post hoc Tukey's test, $p < 0.05$) from the respective nonstimulated group are indicated by *. Values for intact rats are given as reference values and not included in the analysis

#Data adopted from Angelov et al. (2007)

innervated by six longitudinal branches of the buccal branch of the facial nerve (the short common trunk of the fused ramus buccolabialis superior and ramus buccolabialis inferior; Dörfl 1985). There are no proprioceptors in the mystacial musculature (Stal et al. 1987, 1990; Welt and Abbs 1990; Rice et al. 1997; McComas 1998; Whitehead et al. 2005).

Qualitative examination revealed two major features of the m. levator labii superioris. Compared with rats that received FFA + MS, the incidence of intramuscular axonal branches was higher and diameters of muscle fibers were smaller in rats with FFA only, FFA + IONex, as well as FFA + IONex + MS.

Our qualitative observations were matched by quantitative assessments of vibrissal function and the degree of polyinnervation. Thus, rats with FFA, FFA + IONex, and FFA + IONex + MS consistently had poor function and a high percentage of polyinnervated endplates (53 ± 10%, 43 ± 9.4% and 51 ± 10% respectively; Table 2.29). By contrast, rats with FFA + MS had normal vibrissal function and the degree of polyinnervation endplates was significantly smaller (22 ± 5%).

Chapter 3
Discussion

3.1
Significance of Axonal Branching at the Lesion Site

3.1.1
Reduced Collateral Branching Failed to Promote Recovery of Whisking Function

This part of the present synopsis provided, for the first time, controlled experimental evidence for the contribution of axonal branching and misdirection to the failure of recovery of function following facial nerve injury. By manipulating the local environment using neutralizing antibodies to growth factors, we achieved a strong reduction in collateral axonal branching from the proximal stump and a significant improvement of the reinnervation quality in several groups of rats. In the same animals, however, function of the reinnervated vibrissae muscles remained as poor as in nontreated injured animals. As a potential reason for the ineffectiveness of the treatment we identified the well-known posttransectional polyneuronal innervation of the motor endplates, a phenomenon which was not directly manipulated in our experiments. These results raise questions of fundamental importance with regard to the mechanisms limiting functional recovery and to the perspectives for identifying new efficient treatment strategies.

The first issue which has to be addressed is the striking discrepancy between the results of the morphological and the functional assessment of postlesional recovery. This discrepancy cannot be attributed to technical reasons. On the contrary, a number of studies document the reliability and reproducibility of the applied methods. The multiple retrograde labeling technique has been established several years ago and applied in studies aiming to identify effective ways to manipulate axonal branching and misdirection (Angelov et al. 1999; Dohm et al. 2000; Streppel et al. 2002). The sensitivity of the biometric analysis of function has been proven as well (Tomov et al. 2002). Thus, the present results clearly indicate the necessity to study, in any experimental reinnervation paradigm, structure, and function simultaneously. A search in the literature shows that such combined studies are the exception rather than the rule in the field of muscle reinnervation research (Z'Graggen et al. 1998; Merkler et al. 2001; Ramon-Cuéto et al. 2000; Valero-Cabre and Navarro 2002; Li et al. 2003). It appears that our current knowledge on the mechanisms limiting functional recovery is based on analytical comparisons of independently studied structural and functional phenomena that have never been proven to be causatively linked. New experimental approaches have to be devised allowing to evaluate the importance of all putative factors limiting functional recovery.

3.1.1.1
Minor Contribution of "Numerous Regrowing, but Misguided Axons"

Misdirection of regenerating axons to improper targets is caused by collateral branching of the severed axons and nonselective random regrowth of the collaterals along different nerve fascicles (Brushart and Seiler 1987; Nguyen et al. 2002). As a consequence, the myotopic organization of muscle innervation is lost as shown here for the facial nucleus. Individual neurons from a given motor nucleus reestablish contacts with the proper target, with false targets, or both (Table 2.2). The logical conclusion from these anatomical findings is that coordinated muscle function should be impaired: falsely rewired motoneurons will deliver, upon recruitment, impulses to wrong targets and with inappropriate timing. This assumption is widely accepted and axonal misdirection has been identified as the major target for therapeutic manipulations (Brushart and Mesulam 1980; Madison et al. 1999). The results of Sect. 2.1.1 question the fundamental importance of collateral axonal branching as part of the phenomenon of misdirected reinnervation: increased portion of unbranched motoneurons to 70% in treated rats vs. 30% in nontreated animals had no positive influence on function. A reverse structure–function relationship has previously been observed by us in blind rats of the Sprague-Dawley strain using the same experimental approach (Tomov et al. 2002). In these animals, similar to rats with normal vision, the retrograde labeling displayed poor accuracy of target reinnervation and extensive collateral axonal branching. As estimated by the video-based motion analysis, however, recovery of vibrissae movements was near-perfect in the blind rats and poor in the control animal group. Finally, similar discrepancies have been observed by Valero-Cabre and Navarro (2002) who have studied the accuracy of reinnervation and locomotion performance in rats after nerve resection and use of various conduits to reconstruct the sciatic nerve. The conclusion from the existing experimental evidence is that there is no correlation between reduced collateral axonal branching and functional restoration.

3.1.1.2
Major Contribution of "Target Muscle Polyinnervation"

In search for an explanation of the presumption that axonal misdirection may be of secondary importance, one has to consider several aspects of motor unit physiology and plasticity. Normal muscle use requires activation of relatively small proportions of the motoneurons supplying a given muscle (Hennig and Lomo 1985). Also, motoneuron activity is controlled by multiple circuitries which allow different use of a cell pool under normal conditions. For example, both alternating phasic activity of motoneurons innervating antagonistic muscles to enable rhythmic movements in a joint as well as simultaneous activity of the same groups of cells when stabilization in the joint are required. Moreover, the cells in a motor nucleus can be "reeducated" to subserve new functional use as shown in patients after muscle tendon transfer (Illert et al. 1986; Wiedemann et al. 1997; see however Gruart et al. 2003). Finally, the recruitment of single motor units can be modified via descending control mechanisms as indicated by feedback EMG studies in humans (Guntinas-Lichius 2004). On the basis of these considerations, one can speculate that a pool of motoneurons, reduced in size but properly projecting, may be sufficient to control muscle contraction in a physiologically adequate way. The activity of the aberrantly innervating cells may be reduced or modified, by spinal reflex mechanisms and supraspinal control circuitries, so that the functional disturbances due to this abnormality in reinnervation are minimized.

Persistent polyneuronal innervation of muscle fibers has also been considered as a factor limiting recovery (Schröder 1968; Friede and Bischhausen 1980; Gorio et al. 1983; Barry and Ribchester 1995; Grimby et al. 1989; Trojan et al. 1991; Tam and Gordon 2003). Formation of endplates innervated by different neurons on individual muscle fibers is a transient phenomenon during normal development. In contrast, following reinnervation polyneuronal innervation persists for long time periods after establishment of nerve–muscle contacts (Esslen 1960; Mackinnon et al. 1991; Jergovic et al. 2001; Ijkema-Paassen et al. 2002; Grant et al. 2002). It has not been experimentally tested whether and to what degree is this persisting aberrant innervation harmful. General reasoning suggests that the performance of a muscle fiber controlled by two or more asynchronously firing motoneurons cannot be physiologically advantageous. Another form of aberrant reinnervation is the serial approaching and contacting of endplates on different muscle fibers by single axonal collaterals (Son et al. 1996; Trachtenberg and Thompson 1996; Rich and Lichtman 1989). Normally, the wiring pattern within a motor unit is parallel, that is, each terminal axonal branch of a motoneuron supplies one single endplate, which enables synchronous contraction of the muscle fibers. Formation of serial synapses leads to asynchronous contractions evident in EMG recordings from single motor units in patients (Sumner 1990; Montserrat and Benito 1988; Fu and Gordon 1997; Guntinas-Lichius 2004).

We examined one target muscle (LLS) for aberrant reinnervation and observed a qualitatively similar picture in all SD rats: abnormally dense meshwork of intramuscular axonal branches and approximately 50% polyinnervated endplates, a state that can be defined as "hyperneurotization" of the muscles. Of course it would have been by far more convincing if we could provide data on the degree of polyneuronal innervation in the intrinsic follicular muscles. This has never been done before and we faced, unfortunately, serious obstacles. The very delicate slings of the follicular muscles turned out to be extremely variable in structure among different animals and even within one animal. Furthermore, tissue sections of extremely variable appearance were obtained because of the small size of the muscles, the inevitable variability in the planes of sectioning of different muscles, and distortions and displacements during sectioning of tissue blocks which are largely inhomogeneous with respect to the hardness of the tissue components. Thus, any sampling procedure that was considered and attempted turned out to be inadequate for obtaining unbiased, reliable counts in sections from the intrinsic vibrissal musculature.

Hence, we hypothesized that it is the hyperneurotization of muscles and the polyinnervation of their motor endplates which are the cause for the poor functional recovery despite an improved accuracy of reinnervation in animals treated with neutralizing antibodies. This hypothesis could not be tested directly in the corresponding animal groups 2–9, because the zygomatic, buccal, and marginal mandibular branches of the facial nerve were transected and instilled with crystals of the fluorescent tracers DiI, FG, and FB, respectively. Since these dyes are also well-known anterograde neuronal labels, they could compromise our systematic observations on the regrowing intramuscular axons after immunstaining for tubulin. This is why we proved our hypothesis in animals, that were not subjected to triple retrograde labeling and found that the polyneuronal reinnervation of the motor endplates was of decisive importance for the recovery of function: In the poorly performing SD rats, the relationship between mono- and polyinnervated motor endplates was 46–51%, whereas in the well-performing RCS/SD rats, it was 88% vs. 10%. Nevertheless, additional experiments should be designed in which the degree of intramuscular sprouting and polyneuronal innervation can be manipulated, for instance by muscle activity imposed artificially during the phase of synaptic formation and consolidation (Brown et al. 1977; Brown and Holland 1979; Al-Majed et al. 2000; Tam et al. 2001; Brushart et al. 2002; Love et al. 2003).

3.1.2

Effect of Perturbed Microtubule Assembly

A microtubule is a long, hollow cylinder that is made of a polymer of α- and β-tubulins and has a diameter of 25 nm. It has intrinsic polarity, with a fast-growing "plus end" and an opposite, slow-growing "minus end." In axons, microtubules run in a longitudinal orientation and serve as rails along which membranous organelles and macromolecular complexes can be transported; they are unipolar, with the plus end pointing away from the cell body (Hirokawa and Takemura 2005).

With this experimental set, we provided the first evidence for improved restoration of function after peripheral nerve transection by local stabilization of microtubules with 10 μg/ml taxol. This beneficial influence has been most probably achieved through pharmacological reduction of the intramuscular axonal sprouting which in turn diminished significantly the polyinnervation of the motor endplates.

We provided, thus, further experimental evidence for a causal relationship between functional outcome after BBFN transection/repair and the intensity of regrowth in the proximal nerve stump (as determined by the amount of microtubules in the regrowing axons). Better restoration of vibrissae motor performance has been shown to be directly associated with reduced polyinnervation of the motor endplates (Guntinas-Lichius et al. 2005; Angelov et al. 2007). This reduction has been most probably caused by a rapid regrowth of neurites to reach the neuromuscular junction and a rapid withdrawal of the redundant axonal terminals from the motor endplate.

Accumulating knowledge shows that neurite regrowth is part of the neurons' receptor-mediated response to extracellular guidance cues (English 2005). Since most receptor-mediated signal transduction pathways converge onto the Rho-family of small GTPases, axonal elongation is associated with substantial reorganization of the cytoskeleton (McHale et al. 1995; King et al. 2001; Guan and Rao 2003; Hahn et al. 2005). This is why, in our work, we tried to increase the rate of axonal regrowth by altering the dynamics of postlesional cytoskeletal reorganization. There are three major intracellular cytoskeletal components responsible for the cytomechanical forces in the leading edge of the axon: actin microfilaments, myosin, and microtubules (Challacombe et al. 1996). In this report, we concentrated our observations on the microtubules.

3.1.2.1

Impact of Microtubules on Axonal Regrowth

Nowadays, it is generally accepted that the actin-based motility causes the movement of microtubules toward the actin-rich peripheral domains of the growth cones (Santos Da Silva and Dotti 2002; Schaefer et al. 2002; Vignjevic et al. 2003). The retraction of F-actin from the leading edge of the growth cone after cytochalasin treatment (inhibition of actin polymerization) causes a complete loss of the guidance capabilities (Bentley and Torojan-Raymond 1986).

Anyway, microtubules preserve their capability to extend to the leading edge (Marsh and Letourneau 1984; Forscher and Smith 1988; Sahly et al. 2005). Accordingly, numerous earlier studies indicate that tubulin is involved mainly in axonal elongation, a process considered of secondary importance when compared with axonal navigation (Yamada et al. 1970, 1971; Hoffman and Cleveland 1988; Hoffman et al. 1992; Moskowitz et al. 1993).

Recent direct observations, however, show that at axonal branching points, the focal accumulation of F-actin is always accompanied by splaying of looped or bundled microtubules, that is, dynamic microtubules colocalize and copolymerize with F-actin (Gallo and Letourneau 1999; Dent and Kalil 2001). This might be caused by local attenuation of F-actin flow associated with growth cone–target interactions (Suter and Forscher 2000), or dynamic microtubule ends may be actively captured by actin filaments (Bentley and O'Connor 1994). Microtubuli, in turn, facilitate the fusion of vesicles with the plasma membrane, promoting the extension of growth cone lamellipodia (Spira et al. 2003; Kalil and Dent 2005). Upregulated levels of tubulin in the perikarya and increased delivery of mictotubules to regrowing axon tips have been considered essential for regeneration (Tetzlaff et al. 1988a, b, 1991, 1996).

Accordingly, Schaefer et al. (2002) and Fukata et al. (2002) show that the population of microtubules that invade the peripheral domain via filopodia are highly dynamic, suggesting functional specializations, perhaps in exploratory and/or signaling capacity. Finally, drugs that attenuate either microtubule or actin dynamics (inhibition of actin polymerization with cytochalasin, stabilization of microtubules with taxol, or damping of microtubule dynamics with vinblastine) have been shown to inhibit axonal branching but not elongation (Baas and Ahmad 1993; Tanaka et al. 1995; Williamson et al. 1996; Challacombe et al. 1997). Treatment with vincristine, an inhibitor of microtubule formation blocks the outgrowth of some axons and delays the regeneration of others (Pan et al. 2003).

3.1.2.2
Effects of Altered Tubulin Dynamics on Axonal Regrowth, Collateral Axonal Branching, and Quality of Target Reinnervation

Since it has already been shown in vitro that stabilization of microtubules with taxol improves axonal elongation (Challacombe et al. 1997; Baas and Ahmad 1993; Tanaka et al. 1995; Williamson et al. 1996) we tested whether a similar treatment in vivo would also increase the rate of neurite regrowth and improve recovery of muscle function. Such a test is not only highly relevant to everyday clinical practice, but also could be applied very rapidly – some pharmacological agents that affect microtubule dynamics are registered and established drugs for use in human patients.

Unaltered Amounts of Tubulin in BBFN Despite Varying Treatments

Two months after transection and entubulation, there were no significant differences in the amount of neuron-specific beta-tubulin between the BBFN of intact rats, animals who underwent transection and entubulation of BBFN, and the rats treated with agents to perturb the microtubule assembly. At first sight, this finding may be astonishing: upregulated levels of tubulin in the axotomized perikarya and increased delivery of microtubules to regrowing axon tips have been repeatedly described and considered essential for axonal elongation (Hoffman and Lasek 1980; Hoffman and Cleveland 1988; Tetzlaff et al. 1988a; Peeva et al. 2006). The reason for this presumptive discrepancy lies in the length of the posttransectional survival period: definitive increase in the microtubule content is a feature of the regrowing neurites (for BBFN this would be till maximum 14 days postaxotomy), but not of axons 2 months after transection. Accordingly, the effects of the pharmacological agents applied at the transection and entubulation site are no longer evident at 2 months after nerve reconstruction.

Unaltered Degree of Collateral Axonal Branching Despite Varying Treatment

Assessment of collateral axonal branching by means of retrograde neuronal labeling is possible only in the premise that there occurs no neuronal cell death after axotomy in adult rats. Accordingly, earlier data after Nissl-staining of facial motoneurons in paraffin sections, after retrograde labeling with horseradish peroxidase and after immunostaining of vibratome sections for neuron-specific enolase showed no significant alterations in the number of facial motoneurons either at one or at 8 weeks after facial nerve transection and suture (see Angelov et al 2005 for more details).

In line with our recent observations (Guntinas-Lichius et al. 2005; Angelov et al. 2007), the results of the present study question once again the functional significance of collateral axonal branching at the lesion site. In all rats that had been subjected to BBFN transection, retrograde labeling displayed extensive (21–22%) collateral axonal branching (Fig. 2.4b–d).

As estimated by video-based motion analysis, however, recovery of vibrissal movements was improved in the rats of group G and poor to nonexistent in animals of groups B–F (Table 2.10). Lack of whisking recovery after reduced axonal branching has previously been observed by us after facial nerve injury (Tomov et al. 2002). Similarly, use of conduits to reconstruct the sciatic nerve after injury and improve accuracy of reinnervation did not improve locomotor performance in rats (Valero-Cabre and Navarro 2002). The conclusion from the existing experimental evidence is that reduced collateral axonal branching does not lead to better functional restoration. This unexpected finding cannot be explained to date. We can only speculate that, as a result of use-dependent plasticity in the CNS, recruitment of aberrantly innervating motoneurons may be reduced or modified, by spinal reflex mechanisms and supraspinal control circuitries, so that the functional disturbances due to this abnormality in reinnervation are minimized.

Reduced polyinnervation of the motor endplates that was achieved by taxol treatment correlated with and, may at least in part, explain functional restoration.

Strangely enough, this effect of taxol was not evident in the experiments of BBFN transection and subsequent injection of taxol solution directly into the target musculature. The abnormally dense meshwork of intramuscular axonal branches and polyinnervated endplates (approximately 40%; Table 2.11 last three rows) after BBFN transection persisted. Accordingly, recovery of vibrissal motor performance was poor (Table 2.10). One possible explanation could be that the weekly application of agents known to alter microtubule dynamics directly into the target muscles may block axonal elongation to the motor endplates and initiate intensive compensatory intramuscular sprouting.

3.2 Unsuccessful Ways to Reduce Intramuscular Axonal Sprouting in Denervated Muscles

3.2.1 Intraoperative Electrical Stimulation (IOES) Prior to Reconstructive Surgery

In this chapter, we showed in adult rat that brief electrical stimulation immediately after transection and for 1 h prior to end-to-end suture of the severed facial nerve, a purely motor nerve tract, did not lead to improved motor recovery at 4 months. Regardless of whether the animals were electrically stimulated or not, the degree of collateral branching of axons at the lesion site was high (50–70%), the proportion of polyinnervated motor endplates in the musculature was approximately 50% and the amplitude of vibrissal whisking remained at 25–30% of that in intact animals. By contrast, MS over a 4-month period restored near-normal whisking function and significantly reduced the degree of polyinnervation.

In comparison to small laboratory animals, in humans, regeneration often has to occur over long distances; as a consequence, reinnervarion of peripheral targets is slow and requires many months to be accomplished. In addition to slow axon regeneration across the injury site and within distal stumps, there is a progressive decline in the regenerative capacity of axotomized neurons, a declining ability of denervated Schwann cells to support axonal regeneration and worsening muscle atrophy (Brushart et al. 2002; Fenrich and Gordon 2004; Irintchev et al. 1990). Collectively, these factors are considered to be the major hurdles limiting recovery after peripheral nerve injury in humans. Because of the limited window of opportunity for successful regeneration clinically, accelerated reinnervation is supposed to improve the final outcome to a considerably greater extent in humans than in small laboratory animals in which absolute distances faced by regenerating axons are relatively small.

Clinically, there are few options for treating denervated mimic muscles. Although 40 years of ES research have shown a variety of neuromuscular benefits, a great deal of controversy surrounds its use with either some benefit (Farragher et al. 1987; Cole et al. 1991; Williams 1996; Targan et al. 2000; Nicolaidis and Williams 2001; Marqueste et al. 2006) or no effect (Mosforth and Taverner 1958; Huizing et al. 1981; Waxman 1984; Moller and Sen 1990; Kuroki et al. 1994; Ishikawa et al. 1996; Gittins et al. 1999; Marqueste et al. 2002; Diels 2005; Dow et al. 2006) being described. For example, electrical stimulation of denervated soleus muscle inhibits intramuscular sprouting and diminishes motor-endplate polyinnervation (Brown et al. 1980; Love et al. 2003). However, regular ES of totally denervated muscle fibers suppresses the production of chemical mediators required for reconnection of an axon branch with its motor endplate on the muscle and also reduces the spontaneous electrical activity of orphaned muscle fibers (fibrillation), which is thought to be a signal for sprouting of the remaining healthy motor nerve (Cohan and Kater 1986; Brown and Holland 1979). By contrast, ES of muscle fibers that retain a partial nerve supply may simulate voluntary muscle overuse and contribute to suppression of the chemical mediators required for the reinnervation of the denervated fibers (Diels 1995). For the above reasons, ES is not a method of choice and indeed has not been widely used to treat facial paralysis.

However, tantalizing supportive evidence for the benefits of electrical stimulation has shown that, if applied at the time of nerve lesion, ES in rats enhances axonal regrowth, precision of reinnervation (i.e., the preferential motor reinnervation), and expression of GAP-43 by sensory neurons (Brushart et al. 2005; Geremia et al. 2007). In a clinical setting, these effects might lead to advantages in sensory recovery, possibly even alleviation of neuropathic pain, in addition to facilitating the return of motor abilities.

In agreement with the observations of Ahlborn and colleagues (2007) after femoral nerve injury in mice, we have found here that ES does not improve the functional outcome or reduce aberrant regeneration after facial nerve reconstruction in rats. The only positive effect of ES that was detected was a transient improvement, between 1 and 3 months after FFA and ES, of protraction velocity. These findings do not preclude the possibility that ES could nevertheless be more efficient in humans and large animals than in small laboratory rodents. In addition, one has to also consider that the effects of electrical stimulation could depend on the type of nerve to which this treatment is applied. Positive effects observed after ES of the femoral nerve, such as accelerated preferential motor reinnervation and axonal regrowth (Al-Majed et al. 2000; Brushart et al. 2002) and accelerated functional recovery (Ahlborn et al. 2007), might be due to stimulation of both motor and sensory neurons projecting through the distal stump of this mixed nerve. In contrast, electrical stimulation of the facial nerve, which is a purely motor nerve, does not have direct influences on sensory neurons and this might lead to a reduced efficiency as compared with stimulation of a mixed nerve.

In conclusion, while acute ES at the time of injury appears to have some transient benefit in small animal laboratory models following injury to mixed peripheral nerves, we show here that this intervention does not confer any long-term benefit following injury to a purely motor nerve. The transient benefits of ES following mixed peripheral nerve injury contrast with the absence of any functional recovery after application of MS to the forearm following median nerve injury (Sinis et al. 2008). One possibility to explore now might be to combine acute ES with chronic MS to maximize functional recovery after injury to mixed peripheral nerves.

3.2.2
Postoperative Electrical Stimulation (POES) of Paralyzed Vibrissal Muscles

Here, we showed in adult rats that electrical stimulation three times weekly for 2 months starting 1 day after end-to-end suture of the transected facial nerve, a purely motor nerve tract, does not improve motor recovery. Intriguingly, ES reduced the number of innervated motor endplates to 20% of normal values, thus, causing partial muscle reinnervation. Furthermore, regardless of whether animals received ES or were sham stimulated, both collateral axonal branching and the proportion of polyinnervated motor endplates were elevated.

Anyway, it is by no means certain that the demonstrated devastating effect on reinnervation would be replicated in human muscles, which may tempt orthopaedic surgeons and physiotherapists to make unjustified extrapolation from the rat facial nerve model to the human limb muscles. ES is currently applied to counteract the severe muscle atrophy with interstitial fibrosis, which may occur during the very long (6 months or more after brachial plexus surgery) period of reinnervation, that is, maintaining muscle fiber size and structure during the period of nerve regrowth; ES should render the outcome more successful.

The general neurobiological question, whether muscle stimulation during the period of reinnervation would be beneficial is still a major unresolved issue. Earlier findings that electrical stimulation leads to reduced intramuscular axonal sprouting in partially denervated muscles (Tam et al. 2001) have raised concern that muscle reinnervation might be compromised (Eberstein and Eberstein 1996). Indeed, Hennig (1987) has reported diminished degree of reinnervation, but, at the same time, other experiments have shown either positive (Cole and Gardiner 1984; Einsiedel and Luff 1994; Al-Majed et al. 2000; Mendonca et al. 2003; Gordon et al. 2007) or no effects (Herbison et al. 1973).

Despite a lack of sufficient knowledge from animal experiments, ES of muscles has been widely used in human patients as a rehabilitation treatment over decades for a variety of neural injuries (see Sect. 3.2.1). The success of ES appears to depend on the extent of muscle denervation/reinnervation.

By contrast, ES can have adverse effects in partially denervated muscles by stimulating voluntary muscle overuse and suppressing the production of chemical mediators required for reinnervation of denervated muscles (Diels 1995; Tam et al. 2001). In addition, ES of partially denervated muscle reduces the spontaneous electrical activity (fibrillation) of denervated muscle fibers, which is thought to be a signal for sprouting of the remaining healthy motor nerve (Cohan and Kater 1986; Brown and Holland 1979).

In the current study, rather than implanting electrodes, we minimized the invasiveness of the ES procedure by using acupuncture needles, which were placed at some distance from the motor point where the majority of motor endplates are located. We delivered ES sufficient to activate axons but not muscle fibers throughout the period of denervation and reinnervation. We did not show any benefit in return of whisking function; rather, we showed the opposite in that ES drastically reduced the degree of muscle fiber reinnervation. To our knowledge, this is the first direct demonstration that ES reduces motor endplate reinnervation although there are a number of studies that indirectly support our observation. In vitro, ES significantly increases neuromuscular synapse elimination compared to that observed in nonstimulated cultures (Nelson et al. 1993). In vivo, ES of partially denervated muscles has been also shown to have positive effects (see above).

Taken together, the findings suggest that while electrical stimulation might confer benefit in some situations (e.g., denervated muscles of the extremities with larger motor units), it appears to elicit an adverse effect when applied to denervated small and fine muscles of the face.

3.3 Successful Ways to Reduce Intramuscular Axonal Sprouting in Paralyzed Muscles

3.3.1 Manual Stimulation of Paralyzed Vibrissal Muscles After FFA

This report provided the first controlled experimental evidence for the efficacy of mechanical muscle stimulation to improve functional recovery after facial nerve injury in the rat. By stroking the whiskers we stimulated their fine vibrissal muscle slings innervated by the facial nerve and achieved a significant reduction of polyinnervated motor endplates as well as a full recovery of vibrissal motor performance.

Restoration of useful function after peripheral nerve injury is a major challenge for reconstructive surgery and rehabilitation medicine (Lundborg 2003). Recent clinical findings have indicated that "facial retraining" using physical rehabilitation can partially improve outcome in a variety of conditions involving facial nerve injury such as acoustic neuroma, Bell's palsy, Ramsay Hunt syndrome, and facial nerve anastomosis (Barbara et al. 2003; Van Swearingen and Brach 2003).

Restoration of vibrissal whisking in rodents is a useful model to study functional recovery after facial nerve injury in humans. However, there should be no confusion with facial hairs in humans. To avoid any misinterpretation, we stress that by stroking the whiskers, we also stimulated their fine vibrissal muscle slings, which are innervated by the facial nerve. There is no parallel with human facial hairs since arrector pili muscles are absent from human facial hairs, eyelashes and eyebrows, the hairs around the nostrils, and the external auditori meati (Bannister 1995). In addition, arrector pili muscles of hairs elsewhere in the human body are innervated by noradrenergic sympathetic axonal terminals, not peripheral motor axons. Here, we not only show that manual stimulation can restore whisking function after facial nerve injury in rat but also provide evidence for the mechanisms underlying the recovery.

An important and clinically relevant finding in this investigation is that the success of treatment is determined by the type of stimulation. Recovery under enriched environmental conditions stimulating the use of face muscles, in particular those controlling the vibrissae, was completely ineffective. This is surprising given the known stimulating effects of enriched environments on neuronal plasticity and on adaptive responses (Van Praag et al. 2000). Indeed, clinical success has been thought to rely exclusively on plasticity of cortical and subcortical neuronal networks (Sanes and Donoghue 2000). Our results do not necessarily contradict this view. In our experimental paradigm, sensory innervation of the face remains intact and there should be no alteration of the somatosensory cortical representation. Therefore, cortical plasticity may be of primary importance for restoration of sensory, but not motor function (Bisler et al. 2002). The surprisingly high efficacy of the mechanical stimulation of the muscles indicates the importance of applying functionally relevant stimulation protocols (Beazley et al. 2003; Dunlop and Steeves 2003).

The analysis of reinnervation pattern of m. levator labii superioris revealed qualitatively similar picture in all operated rats, which lacked mechanical stimulation: there was an abnormally dense meshwork of intramuscular axonal branches and polyinnervated endplates after facial nerve transection and suture and after transection and suture followed by environmental enrichment (Table 2.15). In contrast, the proportion of polyinnervated endplates was significantly reduced in animals with transection and suture followed by manual stimulation alone or followed by manual stimulation together with environmental enrichment. The effects of mechanical muscle stimulation are readily explainable since previous studies have shown that muscle activity imposed artificially during the phase of synaptic formation and consolidation leads to reduction of the intramuscular sprouting (Brown et al. 1980; Tam et al. 2001; Deschenes et al. 2006).

An earlier study has suggested that intramuscular axonal sprouting in response to muscle paralysis occurs because of short-range diffusible sprouting stimuli generated by the inactive muscle fibers (Brown and Ironton 1977). Accordingly, two subsequent reports have shown that direct muscle, but not nerve stimulation, inhibits intramuscular sprouting possibly by interaxonal competition-induced counteracting and/or neutralization of sprouting stimuli arising from the denervated muscle fibers (Brown and Holland 1979; Love et al. 2003). A reduction in perisynaptic Schwann cell processes bridging innervated and denervated endplates might also have occurred (Tam et al. 2001).

In conclusion, whereas the exact mechanisms linking mechanical stimulation, polyinnervation, and restoration of the muscle function are still unknown, the facial nerve transection model provides a very good system to address this issue in future experiments. The present report provides clear evidence that manual mechanical stimulation of the denervated muscles can "override" the effects of the robust and consistent (Mackinnon et al. 1991; Reynolds and Woolf 1992), but inappropriate (Esslen 1960), axonal regrowth in the target muscles by reducing the degree of polyinnervation. This effect is apparently sufficient for restoration of motor function. Our findings are pertinent to developing rehabilitation strategies for peripheral nerve injury since they suggest that muscle reinnervation, rather than misdirected axonal regrowth should be targeted for therapeutic manipulation.

3.3.2
Manual Stimulation of Paralyzed Facial Muscles After HFA or IPNG

With this chapter, we confirmed our recent experimental evidence for the contribution of manual stimulation to the recovery of vibrissal function following facial nerve injury (Angelov et al. 2007) and now show a positive effect after two common types of facial nerve reconstruction, HFA and IPNG. As for FFA, improved recovery of vibrissal motor performance after HFA and IPNG was associated with a significant reduction in the proportion of polyinnervated motor endplates. Our results provide new perspectives for implementation of efficient and effective clinical treatment strategies.

3.3.2.1
Clinical Relevance of Both Types of Nerve Reconstruction (HFA and IPNG)

When the proximal stump of the facial nerve is not available for anastomosis, surgical transpositions of cranial nerves to the distal facial nerve stump is the only feasible reconstruction approach. Transposition of the hypoglossal nerve and end-to-end anastomosis directly to the facial nerve (HFA) is a frequently used technique (Manni et al. 2001). The main trunk of the facial and hypoglossal nerve on the paralyzed side of the face are cut, and the proximal stump of the hypoglossal nerve anastomosed by nerve suture to the distal stump of the facial nerve. Hypoglossal axons regrow into the facial nerve branches and reinnervate the facial musculature.

In other cases, such as following removal of malignant facial nerve tumors, the distal facial nerve stump is destroyed and interpositional nerve grafting (IPNG) is thus mandatory (Bhathia et al. 1995; Saleh et al. 1995). Another common example is parotid cancer surgery, which often results in damage of several nerve branches distal to their bifurcation from the common facial nerve trunk requiring several nerve grafts. Discrepancies in caliber between the thick proximal stump and the thin distal branches are overcome by pooling the peripheral branches or by splitting the interposition graft (Guntinas-Lichius 2004). Despite these difficulties, IPNG restores some voluntary facial expression and function after resection of parotid neoplasms (Fisch and Lanser 1991).

Generally, recovery of function after HFA or IPNG is poor and voluntary movements in human patients are absent (Manni et al. 2001; Tankere et al. 2003). Patients often complain of subjective oral dysfunction, such as difficulty in masticating, articulating, and swallowing (Kunihiro et al. 2003). Nevertheless, spontaneous, that is, subconscious smiling, can occasionally (especially in young patients or children) be restored indicating a remarkable degree of neuronal plasticity in the brainstem or higher motor centers of man (May 1986).

3.3.2.2

Factors that May Contribute to Improved, but Not Complete, Recovery of Function After HFA and IPNG

Similar to FFA, function following HFA and IPNG in a clinical setting is less than optimal and outcomes in patients can be improved only to a limited extent by postoperative training (Stennert and Limberg 1982). In contrast to FFA, however, our present results show, that despite daily MS, recovery of whisking after HFA or IPNG is not complete. One major reason for this could be that under HFA and IPNG circumstances, regeneration needs more time to reach the regeneration under FFA conditions.

Compared to HFA, IPNG is a severe surgical intervention, involving removal of 5–6 mm of nerve length, two suture sites, more difficult adaptation and coaptation of the stumps, and more intensive scar formation. Whereas the severity of the lesion coupled with more extensive collateral branching presumably leads to poorer functional recovery following IPNG, the situation after HFA is more complicated.

Extensive counts of neurons in adult rat have revealed that practically 100% of the hypoglossal motoneurons survive after HFA (Watson 1965). Earlier work in rat has shown that transected hypoglossal axons readily regrow into the facial "periphery." Neuromyographic recordings from the m. levator labii superioris have shown that the first EMG responses have a nerve conduction velocity (NCV) of 1–2 m/s as early as 21 days after HFA. Thereafter, NCV increases continuously reaching about 50% of the preoperative mean (26.3 ± 1.3 m/s) 8 weeks after HFA. The data suggest that, after HFA, the cut hypoglossal axons are able to reach and reinnervate the whiskerpad muscles. Indeed, the number of hypoglossal motoneurons that project into the muscles of the whiskerpad increases steeply and, 2 months after HFA, reinnervation of the vibrissal muscles is almost complete (Schneider et al. 1994).

Why then is the recovery of function worse after HFA as compared to FFA? A possible explanation might lie in the different physiological behavior of facial and hypoglossal motoneurons. Apart from the tongue, hypoglossal motoneurons also partially supply muscles that dilate the pharynx (genioglossus, geniohyoid, sternohyoid, and sternothyroid muscles). The activity of hypoglossal motoneurons is modulated by respiration and they help maintain upper airway patency and participate in ventilatory homeostasis. The frequency of discharge is 5.2 ± 0.6 times per minute (Morin et al. 1992), that is, about 0.12 Hz. If this frequency is maintained after HFA, all facial muscles with hypoglossal nerve supply will contract every 12 s synchronously in response to respiration-related impulses.

This frequency of impulses, however, is far lower than that of the original facial supply of the whiskerpad, which also lacked rhythmic input. During exploration, rats move their mystacial vibrissae back and forth at about 7 Hz and additionally display a tremor-like movement of the vibrissae at 9 Hz (Semba and Egger 1986). Therefore, the follicular muscles of the whiskerpad (Dörfl 1982), the normal activation of which is closely integrated into the neuronal circuitry of sensory afferents, might not be adequately stimulated by spontaneous hypoglossal motor activity. As a result, underactive muscles might continue to secrete reinnervation-promoting factor(s), which in turn might lead to increased intramuscular branching (Love et al. 2003). Although this explanation may not account for all differences between HFA and FFA (the neuronal pool supplying the muscles that dilate the pharynx is too small), we take it into consideration only for the sake of completeness.

3.3.2.3

Further Reasons for Poor Recovery of Function After Peripheral Nerve Reconstruction

Peripherally, collateral axonal branching at the site of FFA, HFA, or IPNG is a major factor contributing to dysfunction, or lack of function, and was not affected by manual stimulation. This finding is in line with our recent observations (Angelov et al. 2007) and highlights the fact that collateral axonal branching appears to be a robust response with little functional significance. Video-based motion analysis showed that recovery of vibrissal movements was complete in manually stimulated rats after FFA and significantly improved after HFA and IPNG. Poor to nonexistent restoration of function was observed in all nonstimulated groups. We do not yet know the reason for, although we are not surprised by, the somewhat reduced function imparted by MS after HFA and IPNG compared to FFA. However, as described above, we speculate that reduced function is related to reinnervation of whisker pads by "foreign" (i.e., hypoglossal) axons in the case of HFA reconstruction and to the complexity of using multiple grafts in the case of IPNG.

A second major factor contributing to poor recovery is intramuscular axonal sprouting. On reaching a muscle target, regenerating axons undergo additional sprouting to reinnervate many incorrect muscle fibers and thus form new and larger motor units (Son et al. 1996; Gordon et al. 2004). Reinnervation of motor endplates by more than one motoneuron, or "polyinnervation", is maladaptive due to one muscle fiber being controlled by two or more asynchronously, and often functionally different, motoneurons (Brown et al. 1981; Rich and Lichtman 1989).

Another factor is that, following denervation and before reinnervation, profound changes also occur within the muscle itself resulting in profoundly reduced muscle bulk, circulation, and connective tissue which also becomes adherent. In the longer term, after complete denervation, muscle membrane properties also change and become relatively nonresponsive to electrical stimulation (Schwarting et al. 1984). For patients with the possibility of nerve regrowth after complete denervation, it is important to minimize fibrosis within muscle connective tissue so that a potentially movable muscle remains. Reinnervation would then allow reacquisition of contractile proteins that allow muscles to work.

In conclusion, our three clinically relevant facial nerve transection models have provided clear evidence that manual stimulation of denervated muscles can "override" the robust and consistent, but inappropriate, axonal sprouting at the motor endplate by reducing the degree of polyinnervation. The outcome is restoration of motor function that is indistinguishable from normal following FFA (Angelov et al. 2007) and close to normal after HFA and IPNG. Our findings suggest that muscle reinnervation, rather than axonal sprouting or cortical plasticity, should be targeted for therapeutic manipulation following peripheral nerve injury. A major advantage is that such strategies are relatively simple to translate into clinical practice both by the clinician and the patient alike.

3.3.3 Manual Stimulation of Paralyzed Orbicularis Oculi Muscle After FFA

In this part of the present synopsis, we confirmed recent experimental evidence for the efficacy of manual stimulation following facial nerve injury in promoting recovery of vibrissal function (Angelov et al. 2007; Pavlov et al. 2008) by showing that manual stimulation also improved motor recovery of another denervated mimic muscle, the orbicularis oculi muscle (OOM). In addition, as for the whisker-pad muscles, improved eyelid closure was associated with a significant reduction in the proportion of polyinnervated motor endplates. Combined, these "proof of principle" findings have immediate implications for clinical rehabilitation following facial nerve injury.

3.3.3.1 Clinically Abnormal Eye Closure and Its Treatment

One of the most disturbing deficiencies in the course of facial palsy is the blinkless eye. In addition to their static expression, these patients suffer many complications from the inability to protect their eyes, including partially obscured visual fields, epiphora (overflow of teardrops upon the cheek), and/or drying of the eye which may in turn lead to keratitis sicca (inflammation of the cornea), corneal abrasions, and loss of vision (Choi and Raisman 2003).

Clinical interventions vary and include surgical approaches whereby static implants passively assist eyelid closure, or microsurgical manipulation of nerves and muscles dynamically stimulate active eyelid closure (Lavy et al. 2004; Terzis 2005; Botti 2006; Mourits and Vuyk 2006). However, results are usually unsatisfactory and functional recovery often poor (Fisch and Lanser 1991; Lundborg 2003; Guntinas-Lichius 2004; Diels 2005). Physiotherapy offers a less invasive alternative which involves progressive, personally designed programs whereby complete facial paresis is treated during phase 1 by application of warm moist, heated towels, and massage three times per day. Once facial movements begin, the second phase recruits selective muscles by training with functional repetitions of facial movements encouraged by tactile stimulation (Deschenes et al. 2006).

To date, soft tissue massage and small amplitude movements have been shown to promote muscle blood flow and to keep the skin and muscles in optimum condition prior to reinnervation (Hovind and Nielsen 1974; Beurskens 1990; Frach et al. 1992; Beurskens 2005; Coulson 2005). Such treatment is also thought to limit the enlargement of motor units (Diels 2005) which arises because extensive axonal sprouting occurs intramuscularly, thus contributing to an "overflow," or hyperinnervation, of muscle targets (Son et al. 1996; Gordon et al. 2004). In addition to innervation being increased in quantity, the quality also changes with multiple incorrect muscle fibers being controlled by two or more asynchronously, and often functionally different, motoneurons.

The finding that soft tissue massage improves outcome clinically is supported by the current study and our previous results, whereby manual stimulation limits the degree of motor endplate polyinnervation within target muscles (Angelov et al. 2007; Evgenieva et al. 2008; Pavlov et al. 2008). In agreement with previous studies (Connold and Vrbova 1990; Deschenes et al. 2006), we suggest that muscle activity imposed artificially by manual stimulation during the phase of synaptic formation and consolidation reduces intramuscular sprouting and therefore the degree of polyinnervation. Indeed, natural approaches involving "facial retraining" improve outcome in conditions such as acoustic neuroma, Bell's palsy, Ramsay Hunt syndrome, and facial nerve anastomosis (Barbara et al. 2003; Van Swearingen and Brach 2003). Although the mechanism is as yet unknown clinically, experimental evidence is mounting to support the case for an underlying principle whereby soft tissue massage, or manual stimulation, promotes functional motor recovery by improving the accuracy of reinnervation patterns at the synapse.

In conclusion, our clinically relevant models provide clear evidence that manual stimulation of denervated muscles after motor nerve injury can "override" the robust and consistent, but inappropriate, axonal sprouting at the motor endplate by reducing the degree of polyinnervation. Functional recovery can be promoted for a range of behaviors including use of mimic muscles and eyelid closure, after facial nerve injury (Angelov et al. 2007) and swallowing/feeding after hypoglossal nerve damage (Evgenieva et al. 2008). Such strategies are noninvasive and relatively simple to translate into clinical practice both by the clinician and the patient alike.

3.3.4 Manual Stimulation of Paralyzed Suprahyoid–Sublingual Muscles After HHA

3.3.4.1 Clinical Relevance of Peripheral Hypoglossal Nerve Injury

Rapid and accurately adjusted tongue movements are paramount for a wide range of functions including breathing, swallowing, licking/mastication, gaping, gagging, coughing, sneezing, vocalization, and vomiting (Lowe 1981; Sawczuk and Moser 2001; Miller 2002).

Although hypoglossal nerve injury has anecdotally been considered rare, lesions may result from tumors (Keane 1996), trauma (Brennan et al. 1993), tonsillectomy (Sharp et al. 2002), anterior cervical spinal surgery (Sengupta et al. 1999), orotracheal intubation (Rubio-Nazabal et al. 2002), carotid endarterectomy (Lindsay et al. 2003; Cunningham et al. 2004), and use as donor tissue for facial reanimation surgery (Wilson et al. 1994). Unilateral hypoglossal damage is considered clinically to be well tolerated due to preservation of taste- and tactile-sensitivity. Furthermore, despite progressive tongue atrophy, only about 10% of patients report difficulties in chewing, swallowing, and speaking at 6 months follow up; however, between 6 and 12 months after damage, dysarthria and dysphagia may dramatically worsen, dysfunction which is due to an ongoing aberrant reinnervation (Conley and Baker 1979). Similar to the face, the tongue comprises many muscles which, although innervated solely by the hypoglossal nerve, often have antagonistic actions (Wilson et al. 1994). Indeed, in laboratory animals, surgical hypoglossal nerve repair did not result in functional recovery due to aberrant axon regrowth and a failure to reach appropriate target muscles (Hosemann et al. 1990).

3.3.4.2
Possible Mechanisms Underlying Improved Function Following Manual Stimulation of the Paralyzed Suprahyoid–Sublingual Muscles After HHA

Mechanisms limiting functional recovery after peripheral nerve injury are poorly understood. Our model provides unique opportunities to investigate the influence of MS on both structure and function.

Alterations in the Cortical Representation

Cortical reorganization has been reported previously after nerve injury (Horvath et al. 2005) and plasticity within cortical and subcortical networks is thought to be involved in clinical examples of muscle reanimation (Sanes and Donoghue 2000). One reason for finding no changes in cortical representation is possibly that we examined animals at 2 months, that is, once target reinnervation had been completed. Examination of animals at earlier stages would determine whether cortical representation was altered during reinnervation. Another possibility is that, in our HHA model, sensory innervation of the tongue-musculature remains intact with no alteration in somatosensory cortical representation. The lack of quantitative changes in the motor cortical representation regardless of MS supports the notion that cortical plasticity may be of primary importance for restoration of primarily sensory, but not motor, function (Bisler et al. 2002).

Motoneuron Size

The finding of larger retrogradely labeled motoneuron perikarya in stimulated than in nonstimulated rats indicates an effect of the MS on regenerated motoneurons. It is possible that MS alleviates axotomy-induced motoneuron atrophy during the recovery period. Alternatively, prevention of atrophy may result from MS inducing a vigorous regenerative response to the second axotomy performed for retrograde labeling (McPhail et al. 2004b). We suggest that the larger cell bodies of stimulated motoneurons indicate a better functional state. Indeed, there is a correlation between degree of recovery and soma size of retrogradely labeled motoneurons reinnervating the quadriceps muscle 3 months after transection and repair of the femoral nerve (Simova et al. 2006). Although we do not know the mechanism whereby motoneuron size is enhanced, we speculate that MS positively influences regenerating motoneurons via enhanced sensory input (see below).

Synaptic Input to the Hypoglossal Motoneurons

Synaptophysin is an integral membrane-glycoprotein of small presynaptic vesicles and neuroendocrine granules (Wiedenmann and Franke 1985). Staining with synaptophysin antibody detects only presynaptic terminals filled with small vesicles but not depleted boutons. We adopted this immunohistochemical approach in our model to examine the well-established rapid detachment of synaptic terminals from motoneurons following nerve injury, a synaptic stripping that is nevertheless reversible. We are, thus, confident that our results show a decline in sensory input to the facial motor nucleus after facial nerve injury and that manual stimulation restores such input to normal levels.

Motor Endplate Reinnervation

The effect of MS that we show here for hypoglossal and following facial nerve (Angelov et al. 2007) injury can be explained by previous studies showing that imposing muscle activity artificially during synaptic formation and consolidation leads to reduction of intramuscular sprouting (Brown et al. 1980).

Sensory Input

In the nerve lesion paradigm used here, motor axons were lesioned but circuitry conveying sensory information from the tongue to the hypoglossal motoneurons via the trigeminal (V), the glossopharyngeal (IX), and the superior laryngeal, that is, the vagus (X) nerve, remained intact (Lowe 1981). One possibility is that MS resulted in recovery of tongue position via enhanced sensory input. After complete spinal cord transection, in which both motor and sensory functions are lost, motoneurons distal to the injury undergo atrophy, their dendritic trees shrink, and become partially deafferented (Kitzman 2005). Cell atrophy and deafferentation of motoneurons is also observed after peripheral axotomy but, in contrast to spinal cord injury, these changes could be reversed after target reinnervation (Sumner and Watson 1971; McPhail et al. 2004a). We speculate that increased sensory input in stimulated animals may aid the regenerative response of the injured motoneurons via, for example, stimulating plasticity in the brainstem.

In conclusion, whereas the precise mechanisms linking MS, polyinnervation, and restoration of the muscle function are still unknown, hypoglossal nerve injury provides a further clinically relevant model to address this issue. Here, we showed that mechanical stimulation of denervated extrinsic suprahyoid–sublingual region and intrinsic suprahyoid and sublingual muscles (*m. longitudinalis sup., m. longitudinalis inf. m. transversus and m. verticalis*) can at least partially "override" the negative effects of extensive but inappropriate axonal regrowth in target muscles. The end result is a reduction in the degree of polyinnervation which in turn significantly improves motor function of the tongue. Our findings have implications for rehabilitation strategies following peripheral nerve injury involving only motor axons since they suggest that therapies should be directed toward muscle reinnervation, that is, within peripheral target tissue, rather than misdirected axonal regrowth at the site of nerve injury.

3.4 Unsuccessful Manual Stimulation of Paralyzed Forearm Muscles After MMA

We showed in this chapter that daily MS of the forearm muscles after surgical reconstruction of rat median nerve did not improve postoperative recovery of function. Regardless of the postoperative treatment, grip force remained at 60–65% of that in intact control animals. In addition, MS did not influence the degree of axonal sprouting nor the extent of polyinnervation of motor endplates.

3.4.1 Clinical Relevance of Median Nerve Injury

The median nerve is one of the most important nerves supplying the upper extremity with both motor and sensory function (Rathakrishnan et al. 2007; Dillon et al. 2007). Within the hand, its branches innervate the thumb, second, third, and the ulnar part of the fourth digit. Following transection, loss of appropriate sensory feedback results in an inability to detect cold and heat leading into unintended, and often severe, thermal injuries (Tsuboya et al. 2007; Thonnard et al. 1999). Furthermore, the median nerve provides motor supply to the intrinsic muscles of the thumb; transection therefore results in loss of thumb apposition and a substantial overall functional loss in the affected hand (Lowe and Freivalds 1999, Nowak and Hermsdorfer 2003).

3.4.2 The Effects of Manual Stimulation

Although injuries involving mixed nerves, such as the median, should intuitively be treated by therapies which target both motor and sensory components, it has yet to be determined whether manual stimulation of muscles, directed primarily at the motor component, would have any benefit.

In the facial/hypoglossal nerve lesion paradigms in which manual stimulation clearly improved recovery of function (Angelov et al. 2007; Evgenieva et al. 2008) only motor axons were lesioned. Thus, the circuitry conveying sensory information from the face/tongue to the facial/hypoglossal motoneurons via the trigeminal (V), the glossopharyngeal (IX), and the vagus (X) nerves (Lowe 1981; O'Reilly and Fitzgerald 1990; Tolu et al. 1993) remained intact. One possibility in this example, where only motor axons are damaged, is that MS resulted in recovery of mimic/ tongue function via enhanced sensory input.

Accordingly, we show here that MS of the forearm following injury to the mixed median nerve does not improve functional recovery and concomitantly does not increase the proportion of monoinnervated endplates within the flexor muscles. Taken together, the findings suggest that a shift toward monoinnervated endplates is somehow prevented by sensory damage following injury to a mixed nerve. Because sensory retraining paradigms for rats have yet to be developed, it is not known whether sensory retraining alone, or sensory retraining combined with MS, would permit an increase in the proportion of monoinnervated endplates and therefore improve function.

Thus, we have reached the question about the mechanism (i.e., cellular and molecular correlates) by means of which MS could have reduced intramuscular axonal sprouting and polyinnervation of the motor endplates. The most probable candidate for cellular correlates are the terminal Schwann cells (TSC). Upon denervation, they can enlarge and sprout processes (bridges) which reach adjacent innervated motor endplates. Using these bridges TSC reach, attract, and direct intramuscular axonal sprouts toward denervated endplates (Son et al. 1996). Interestingly enough, it has been shown that the outgrowth of TSC processes precedes the outgrowth of sprouts from the intact intramuscular axons, that is, TSC are able to initiate intramuscular axonal sprouting (Dickens et al. 2003). The beneficial effect of mechanical stimulation on muscle reinnervation may thus be mediated by interfering with the extension of TSC processes and their ability to bridge between the endplates. Similar results about perturbed formation of TSC bridges, though after running exercise (Tam and Gordon 2003) or electrical stimulation (Love et al. 2003), have been recently reported: any form of artificially excited muscular activity may inhibit the bridge formation by TSC and reduce postlesional intramuscular sprouting.

As possible molecular correlates we may consider short-range diffusible sprouting stimuli (Pockett and Slack 1982) that have been produced by denervated muscles (Cangiano et al. 1984). Various growth factors have been identified as possible candidates for this role. Their amount is inversely proportional to muscle activity (Brown et al. 1981).

Anyway, it is hard to believe that MS could have had an effect only on denervated muscles. As shown by the poor recovery after deafferentation, there must be additional (neuronal) correlates that might govern the initiation of intramuscular axonal sprouting by TSC. In agreement with previous work (Sulaiman et al. 2002), one possible reason for the failure of MS to improve functional recovery in the present study was that the peripheral axons of the DRGs (C_5-Th_1) were transected during surgery. Despite subsequent regrowth, their excessive collateral branching, as evidenced by double-labeled DRG cells, presumably prevented them from conveying accurate sensory information to their target motor neurons in the ventral horn, therefore, compromising motor function further.

The lack of larger retrogradely labeled motoneuron perikarya in stimulated rats may indicate that MS has no effect on regenerated motoneurons. Irrespective of the reason, we can interpret larger cell body size of stimulated motoneurons as an indication for their better functional state. In support of this interpretation is the finding of a correlation between degree of recovery and mean soma area of retrogradely labeled motoneurons reinnervating the quadriceps muscle 3 months after transection and repair of the femoral nerve (Simova et al. 2006). While a link between better motor function and better functional state of regenerated motoneurons appears plausible, the way in which the effect on motoneurons could be achieved is unknown. Here we can again speculate that MS has a positive influence on regenerating motoneurons via enhanced sensory input.

In conclusion, whereas the precise mechanisms linking MS, monoinnervation, and restoration of muscle function are still unknown, median nerve injury provides a model to address this issue in the context of sensory damage. The present report provided clear evidence that MS alone of denervated muscles following injury to a mixed nerve is insufficient to restore function. Our findings suggest that correction of sensory dysfunction will be key for restoring appropriate muscle reinnervation and function.

3.4.3

Significance of the Intact Trigeminal Sensory Input

In this part of the present synpopsis we showed that, following facial nerve repair, brief but persistent manual stimulation of vibrissal muscles prevented the injury-induced loss, as estimated by synaptophysin immunohistochemistry, of afferent synaptic input to facial motoneurons. In addition, following facial nerve injury, there appeared to be a hierarchy of recovery depending on whether sensory input through the trigeminal nerve is intact and whether MS is used. The best recovery was found if MS is applied when the sensory system was intact. However, if sensory input is damaged – the outcome was significantly *worse*. Taken together, our findings suggest that rehabilitation strategies must be carefully tailored according to the extent of both the motor and sensory deficit.

3.4.3.1

Mechanical Stimulation Prevents Deafferentation of Regenerated Facial Motoneurons

Following facial nerve injury, synaptic terminals rapidly detach from motoneurons, a phenomenon well-known as "synaptic stripping" (Blinzinger and Kreutzberg 1968; Graeber and Kreutzberg 1988). This posttraumatic deafferentation is reversible if target reinnervation occurs (Neiss et al. 1992; Guntinas-Lichius et al. 1994; Mader et al. 2004). Quantitative electron microscopic analysis of regenerated cat gastrocnemius motoneurons has, however, revealed that restoration of synaptic inputs is incomplete in several respects (Brännström and Kellerth 1999). Thus, for example, total synaptic frequency (number per unit membrane length) and total synaptic coverage (percent of membrane length covered by synapses) estimated for motoneuron cell somata and proximal, intermediate, and distal dendritic segments recover to 60–81% and 28–48% of normal, respectively.

Here we provided for the first time quantitative evidence for deficient recovery of synaptic inputs to regenerated motoneurons in the facial nerve injury paradigm: First, synaptophysin immunohistochemistry combined with gray value-based densitometry is a well-established approach for estimation of presynaptic terminal densities in different brain regions (see, for example, Masliah et al. 1990; Svensson and Aldskogius 1993; Calhoun et al. 1996; Spiwoks-Becker et al. 2001; Tiraihi and Rezaie 2004). As applied here, this approach provides estimates proportional to numbers of immunopositive structures rather than intensity of immunofluorescence. Second, the facial nucleus contains virtually no interneurons and gamma-motoneurons (Sherwood 2005). Thus, all synaptic terminals in this nucleus make synapses on the dendrites and cell bodies of facial alpha-motoneurons, with the exception of a presumably limited number of axo-axonic synapses. Third, the measurements were performed in the lateral facial subnucleus which contains only motoneurons projecting through the buccal branch of the facial nerve that are all axotomized during FFA. Since essentially all these motoneurons survive axotomy and reinnervate peripheral muscles (Moran and Graeber 2004; Raivich and Makwana 2007), alterations in synaptic inputs cannot be associated with cell survival or success of regeneration.

As any other method for estimation of synaptic inputs, our approach has its disadvantages. In particular, use of this method does not allow assessment of changes in synaptic inputs to different compartments of the motoneuron (cell body and different segments of the dendritic tree). Also, it is impossible to differentiate changes in specific transmitter systems, excitatory, inhibitory and modulatory, influencing motoneuron functions in different ways. In view of these considerations, it is important to note that inhibitory and excitatory synaptic inputs may be differentially affected. Brännström and Kellerth (1999) have observed, for instance, that while the overall restoration of synaptic inputs to spinal motoneurons is deficient, for some types of synapses, such as S-type (presumably excitatory) boutons on proximal dendrites, overcompensation occurs. Irrespective of these remarks, we consider the provided evidence for deficient afferent input to regenerated facial motoneurons of substantial importance, not least because, as indicated by the observations made on animals with MS, there is a possible link between level of synaptic input and degree of functional recovery (see below). Further quantitative studies on regeneration-related changes in specific transmitter systems and cell circuitries are required to gain deeper insights into the relationship between structural synaptic plasticity and functional outcome after femoral nerve repair.

Furthermore, our results on synaptic densities suggest that MS prevents the FFA-related loss of facial nucleus afferents. Since both degree of functional recovery and axon terminal densities were superior in treated than in control rats without MS, it appears that level of synaptic input is a second structural parameter, after degree of polyneuronal muscle fiber reinnervation, which parallels the degrees of functional restoration after FFA. To postulate such a structure–function link is tempting but further studies are required, as indicated above, to verify and understand such correlations. In addition, the explanation of the mechanism underlying the observed effect of MS is not simple. We can speculate that loss and incomplete restoration of facial nucleus innervation without MS is primarily due to loss of input from last order interneurons in the brain stem, including the principal trigeminal nucleus. With MS, the sensory input conveyed directly to these neurons via the trigeminal nerve is enhanced, which in turn leads to augmented loss and/or better restoration of the afferent input to the facial nucleus. The anatomical substrate mediating the effect would then be the vibrissal trigeminal loop, that is, a chain of neurons in the trigeminal ganglion, principle sensory trigeminal nucleus, and subcortical central whisking pattern generator (Kleinfeld et al. 1999; Gao et al. 2001; Kis et al. 2004; Nguyen and Kleinfeld 2005; Leiser and Moxon 2007) interconnected via direct or indirect intrafascicular trigemino-facial brainstem projections. Indeed, there is extensive anatomical, electrophysiological, and clinical evidence for involvement of the trigeminal system in generation of facial muscle responses and blink reflexes (Moller and Sen 1990; Valls-Sole et al. 1992; Zerari-Mailly et al. 2001; Hattox et al. 2002). However, this simple scenario remains questionable regarding the diversity of different projections to the facial nucleus including connections with the neocortex, other nuclei of cranial nerves and the reticular formation (Dauvergne et al. 2001; Popratiloff et al. 2001). Not least, we should consider the theoretical possibility that local (intramuscular) effects of MS on muscle fibers and Schwann cells (see below) alter the retrograde signaling in regenerating motoneurons that favor stabilization of synaptic contact after axotomy.

3.4.3.2
Adverse Effect of Trigeminal Nerve Ablation on Functional Recovery After FFA

One of our main findings was that the quality of motor reinnervation of whisker pads and whisker function was worse when sensory input is damaged. One possible explanation is that, ablating sensory input from the vibrissal whisker pads permanently deprives motoneurons of trophic support, thereby severely limiting their recovery. Indeed, in the spinal cord, complete transection results in simultaneous loss of both motor and sensory function. Motoneurons distal to the injury become partially deafferented, undergo atrophy, and their dendritic trees shrink, a condition that persists due to lack of spontaneous regeneration (Spruston et al. 1995; Segev 1998; Vetter et al. 2001). Deafferentation and atrophy of motoneurons also occurs after peripheral nerve axotomy but, in contrast to spinal cord injury, peripheral changes are reversed following target reinnervation particularly if sensory damage is avoided (Blinzinger and Kreutzberg 1968; Sumner and Watson 1971; Standler and Bernstein 1982; Brännström et al. 1992a, b; Brännström and Kellerth 1998, 1999; Van den Noven et al. 1993).

3.4.3.3
The Effect of Manual Stimulation Depends on the Integrity of the Trigeminal Sensory System

Clinically, soft tissue massage following facial nerve damage has been shown to result in improved blood flow, facial symmetry, and smiling (see Sect. 3.3.3.1). The question raised by our current work in rat is: Why does manual stimulation of vibrissal muscles improve recovery after facial nerve injury if the sensory system is intact but make it worse in the absence of normal sensory input?

When the sensory system is intact, MS must provide near normal sensory input to the facial motorneurons, as well as to the sensory and motor cortex, thereby maintaining appropriate levels of excitability within the facial-trigeminal loop (Sosnik et al. 2001; Minnery and Simons 2003). Such relative stability presumably allows MS to directly benefit the motorneurons themselves. For example, MS could directly affect the denervated muscle fibers by increasing the circulation, reducing fibrosis, and maintaining membrane properties and therefore responsiveness to action potentials once resinnervation has occurred (Schwarting et al. 1984). Since inactive muscles produce abnormally high levels of growth factors, muscles receiving MS may synthesize fewer growth factors and so limit inappropriate intramuscular axonal sprouting (Tam et al. 2001; Love et al. 2003).

Another possible substrate is the TSC which, after axotomy, migrates from the perineurium, enlarges and sprouts "bridges" which reach adjacent innervated motor endplates. Such TSC bridges attract intramuscular sprouts from intact axons toward denervated endplates (Son et al. 1996). Interestingly, TSC processes precede sprouting from the intact intramuscular axons and are, thus, able to initiate intramuscular axonal sprouting (Dickens et al. 2003). Manual stimulation may limit the extension of TSC processes and their ability to bridge motor endplates. Indeed, both running and electrical stimulation perturb TSC bridge formation (Tam and Gordon 2003; Love et al. 2003).

Another mechanism underpinning improved outcome after MS is that the sensory input may aid motoneuron regeneration by stimulating plasticity in the spinal cord, a phenomenon which also appears crucial for successful functional recovery after peripheral nerve injury (Sulaiman et al. 2002; Guntinas-Lichius et al. 2005; Galtrey et al. 2007).

However, as we show, MS after damage to both the facial and infraorbital nerve leads to an even worse outcome compared to when the sensory system remains intact. Given the extreme disturbances in sensory, as well as motoneuronal excitability following damage to sensory afferents (Devor et al. 1989; Schwarz et al. 1983; Spielmann et al. 1983; Bowe et al. 1985) as well as excessive sprouting (Shaw and Bray 1977; Diamond et al. 1987) and loss of GABA-ergic inhibitory control (Castro-Lopes et al. 1993), it is not surprising that recovery was so poor in animals with both facial and infraorbital nerve injury. We speculate that MS "overloads" a system that has already been rendered susceptible to hyperexcitability, possibly leading to excitotoxicity and, thus, irretrievable damage.

Our combined facial and trigeminal nerve injury model provides an opportunity to distinguish the response of motor and/or sensory nerves to injury. The current study clearly suggests that, following damage to both motor and sensory nerves, whether separate or mixed, the priority will be to ameliorate sensory damage. Furthermore, strategies for sensory and motor repair will most likely need to be different.

Chapter 4
Conclusions

1. Treatment of the proximal stump of transected rat facial nerve with antibodies against NGF, BDNF, bFGF, IGF-I, CNTF, or GDNF increased the precision of reinnervation, as evaluated by multiple retrograde labeling of motoneurons, more than two times as compared to control animals. However, biometric analysis of vibrissae movements did not show positive effects on functional recovery suggesting that polyneuronal reinnervation of muscles – rather than collateral branching – may be the critical limiting factor for restoration of function.
2. Application of established pharmacological agents to perturb microtubule assembly toward stabilization (enhanced polymerization with 10 μg/ml taxol) to the transected buccal branch of the rat facial nerve reduced intramuscular axonal sprouting and polyinnervation of the motor endplates which was accompanied by improved restoration of function.
3. Therapy with brief, low-frequency intraoperative electrical stimulation (IOES; 20 Hz) in adult rat immediately after transection and for 1 h prior to end-to-end suture of the severed facial nerve did not lead to improved motor recovery: the degree of collateral branching of axons at the lesion site was high (50–70%), the proportion of polyinnervated motor endplates in the musculature was approximately 50% and the amplitude of vibrissal whisking remained at 25–30% of that in intact animals.
4. Postoperative electrical stimulation (square 0.1 ms pulses at 5 Hz at an ex tempore established threshold amplitude of between 3.0 and 5.0 V) to the vibrissal muscles for 5 min a day 3 times a week did not reduce collateral branching of axons at the lesion site and did not improve functional outcome. Astonishingly, it reduced the number of innervated motor endplates to approximately one-fifth of normal values and failed to reduce the proportion of polyinnervated motor endplates.
5. Postoperative therapy with manual mechanical stimulation of denervated vibrissal muscles after transection and suture of the rat facial nerve (FFA) resulted in a complete return of normal vibrissal motor performance with a concomitant pronounced reduction in polyinnervation of the motor endplates.

6. Postoperative therapy with manual mechanical stimulation of denervated vibrissal muscles after two other commonly used surgical methods of clinical facial nerve reconstruction namely hypoglossal–facial anastomosis (HFA) and interpositional nerve grafting (IPNG) did not completely restore function but, nevertheless, significantly improved the amplitude of whisker movements by 50% compared with untreated animals. Functional improvement was associated again with a reduction in the proportion of polyinnervated endplates.
7. Postoperative therapy with manual mechanical stimulation of denervated orbicularis oculi muscle (OOM) after transection and suture of the rat facial nerve (FFA) improved eyelid closure which was associated with a significant reduction in the proportion of polyinnervated motor endplates.
8. Postoperative therapy with manual mechanical stimulation of denervated sublingual–suprahyoid muscles improved functional recovery (less expressed tongue deviation from the midline) which was associated with a restoration of the total synaptic input to the hypoglossal perikarya and reduced proportion of polyinnervated NMJ.
9. Postoperative therapy with manual mechanical stimulation of denervated rat forearm muscles following transection and suture of the mixed (sensory and motor) median nerve (median–median anastomosis, MMA) did not improve the level of functional recovery, measured by the force of grip. Also, we found no differences among the groups in the degree of axonal sprouting, the extent of motor endplate polyinnervation, and in the soma size of regenerated motoneurons.
10. Mechanical stimulation of denervated muscles exerts its beneficial effect(s) in two possible ways, which do not exclude each other:
    (a) It activates the ipsilateral trigeminal afferents which restore (directly or indirectly) the synaptic coverage of the facial motoneuronal perikarya in the brainstem
    (b) It simulates contractions of the denervated muscles and suppresses the production of excessively large amounts of trophic factors – this, in turn, prevents the immobilization-induced vigorous sprouting of the terminal Schwann cells

References

Ahlborn P, Schachner M, Irintchev A (2007) One hour electrical stimulation accelerates functional recovery after femoral nerve repair. Exp Neurol 208:137–144

Aldes LD (1995) Subcompartmental organization of the ventral (protrusor) compartment in the hypoglossal nucleus of the rat. J Comp Neurol 1995(353):89–108

Aldskogius H, Thomander L (1986) Selective reinnervation of somatotopically appropriate muscles after facial nerve transection and regeneration in the neonatal rat. Brain Res 375: 126–134

Aletta JM, Greene LA (1988) Growth cone configuration and advance: a time-lapse study using video-enhanced differential interference contrast microscopy. J Neurosci 8:1425–1435

Al-Majed AA, Neumann CM, Brushart TM, Gordon T (2000) Brief electrical stimulation promotes the speed and accuracy of motor axonal regeneration. J Neurosci 20:2602–2608

Al-Majed AA, Tam SL, Gordon T (2004) Electrical stimulation accelerates and enhances expression of regeneration-associated genes in regenerating rat femoral motoneurons. Cell Mol Neurobiol 24:379–402

Angelov DN, Gunkel A, Stennert E, Neiss WF (1993) Recovery of original nerve supply after hypoglossal-facial anastomosis causes permanent motor hyperinnervation of the whisker-pad muscles in the rat. J Comp Neurol 338:214–224

Angelov DN, Neiss WF, Gunkel A, Guntinas-Lichius O, Stennert E (1994) Axotomy induces intranuclear immunolocalization of neuron-specific enolase (NSE) in facial and hypoglossal neurons of the rat. J Neurocytol 23:218–233

Angelov DN, Neiss WF, Streppel M, Andermahr J, Mader K, Stennert E (1996) Nimodipine accelerates axonal sprouting after surgical repair of rat facial nerve. J Neurosci 16:1041–1048

Angelov DN, Skouras E, Guntinas-Lichius O, Streppel M, Popratiloff A, Walther M, Klein J, Stennert E, Neiss WF (1999) Contralateral trigeminal nerve lesion reduces polyneuronal muscle innervation after facial nerve repair in rats. Eur J Neurosci 11:1369–1378

Angelov DN, Guntinas-Lichius O, Wewetzer K, Neiss WF, Streppel M (2005) Axonal branching and recovery of coordinated muscle activity after transection of the facial nerve in adult rats. Adv Anat Embryol Cell Biol 180:1–130

Angelov DN, Ceynowa M, Guntinas-Lichius O, Streppel M, Grosheva M, Kiryakova SI, Skouras E, Maegele M, Irintchev AP, Neiss WF, Sinis N, Alvanou A, Dunlop SA (2007) Mechanical stimulation of paralyzed vibrissal muscles following facial nerve injury in adult rat promotes full recovery of whisking. Neurobiol Dis 26:229–242

Anonsen CK, Trachy RE, Hibbert J, Cummings CW (1986) Assessment of facial reinnervation by use of chronic electromyographic monitoring. Otolaryngol Head Neck Surg 94:32–36

Arvidsson J (1982) Somatotopic organization of vibrissae afferents in the trigeminal sensory nuclei of the rat studied by transganglionic transport of HRP. J Comp Neurol 211:84–92

Ashley Z, Salmons S, Boncompagni S, Protasi F, Russold MF, Lanmuller H, Mayr W, Sutherland H, Jarvis JC (2007) Effects of chronic electrical stimulation on long-term denervated muscles of the rabbit hind limb. J Muscle Res Cell Motil 28:203–217

Ashley Z, Sutherland H, Russold MF, Lanmuller H, Mayr W, Jarvis JC, Salmons S (2008) Therapeutic stimulation of denervated muscles: the influence of pattern. Muscle Nerve 38:875–886

Baas PW (1997) Microtubules and axonal growth. Curr Opin Cell Biol 9:29–36

Baas PW (1999) Microtubules and neuronal polarity: lessons learned from mitosis. Neuron 22:23–31

Baas PW, Ahmad FJ (1993) The transport properties of axonal microtubules establish their polarity orientation. J Cell Biol 120:1427–1437

Baas PW, Deitch JS, Black MM, Banker GA (1988) Polarity orientation of microtubules in hippocampal neurons: uniformity in the axon and nonuniformity in the dendrite. Proc Natl Acad Sci USA 85:8335–8339

Baker RS, Stava MW, Nelson KR, May PJ, Huffman MD, Porter JD (1994) Aberrant reinnervation of facial musculature in a subhuman primate: a correlative analysis of eyelid kinematics, muscle synkinesis, and motoneuron localization. Neurology 44:2165–2173

Bannister LH (1995) The Integumental system. In: Bannister LH, Berry MM, Collins P, Dyson M, Dussek JE, Ferguson MWJ (eds) Gray's anatomy, 38th edn. Churchill Liningstone, New York, p 407

Barbara M, Monini S, Buffoni A, Cordier A, Rochetti F, Harguindey A, Di Stadio A, Cerruto R, Filipo R (2003) Early rehabilitation of facial nerve deficits after acoustic neuroma surgery. Acta Otolaryngol 123:932–935

Bardosi A, Goebel HH, Stennert E (1987) The ultrastructure of normal and denrvated human facial muscles. Plast Reconstruct Surg 79:171–176

Barry JA, Ribchester RR (1995) Persistent polyneuronal innervation in partially denervated rat muscle after reinnervation and recovery from prolonged nerve conduction block. J Neurosci 15:6327–6339

Bastmeyer M, O'Leary DDM (1996) Dynamics of target recognition by interstitial axon branching along developing cortical axons. J Neurosci 16:1450–1459

Beazley LD, Rodger J, Chen P, Tee LBG, Stirling RV, Taylor AL, Dunlop SA (2003) Training on a visual task improves the outcome of optic nerve regeneration. J Neurotrauma 20: 1263–1270

Bedi K, Winter J, Berry M, Cohen J (1992) Adult rat dorsal root ganglion neurons extent neurites on predegenerated but not on normal peripheral nerves in vitro. Eur J Neurosci 4:193–200

Bentley D, O'Connor TP (1994) Cytoskeletal events in growth cone steering. Curr Opin Neurobiol 4:43–48

Bentley D, Torojan-Raymond A (1986) Disoriented pathfinding by pioneer neurone growth cones deprived of filopodia by cytochalasin treatment. Nature 323:712–715

Bento RF, Miniti A (1993) Anastomosis of the intratemporal facial nerve using fibrin tissue adhesive. Ear Nose Throat J 72:663

Berg RW, Kleinfeld D (2003) Rhythmic whisking by rat: retraction as well as protraction of the vibrissae is under active muscular control. J Neurophysiol 89:104–117

Bermejo R, Harvey M, Gao P, Zeigler HP (1996) Conditioned whisking in the rat. Somatosens Mot Res 13:225–233

Bertelli JA, Mira JC (1995) The grasping test: a simple behavioral method for objective quantitative assessment of peripheral nerve regeneration in the rat. J Neurosci Methods 58:151–155

Beurskens CHG (1990) The functional rehabilitation of facial muscles and facial expression. In: Castro D (ed) Facial nerve, Proceedings of the Sixth International Symposium on the Facial Nerve. Kugler Publications, Amsterdam, pp 509–511

Beurskens CHG (2005) Mime therapy: contents and evidence. In: Beurskens CHG, van Gelder RS, Heymans PG, Manni JJ, Nicolai JPA (eds) The facial palsies. Complementary approaches. Lemma Publishers, Utrecht, The Netherlands, pp 285–294

Bhathia S, Karmarkar S, Calabrese V (1995) Intratemporal hemangiomas involving the facial nerve: diagnosis and management. Skull Base Surg 5:227–232

Bisby MA (1995) Regeneration of peripheral nervous system axons. In: Waxman SG, Kocsis JD, Stys PK (eds) The Axon. Oxford University Press, Inc., New York, pp 553–578

Bischoff A, Grosheva M, Irintchev A, Skouras E, Kaidoglou K, Michael J, Angelova S, Kuerten S, Sinis N, Dunlop SA, Angelov DN (2009) Manual stimulation of the orbicularis oculi muscle improves eyelid closure after facial nerve injury in adult rats. Muscle & Nerve 39:197–205

Bisler S, Schleicher A, Gass P, Stehle JH, Zilles K, Staiger JF (2002) Expression of c-Fos, ICER, Krox-24 and JunB in the whisker-to-barrel pathway of rats: time course of induction upon whisker stimulation by tactile exploration of an enriched environment. J Chem Neuroanat 23:187–198

Bixby JL, Harris WA (1991) Molecular mechanisms of axon growth and guidance. Annu Rev Cell Biol 7:117–159

Blanco JE, Anderson KD, Steward O (2007) Recovery of forepaw gripping ability and reorganization of cortical motor control following cervical spinal cord injuries in mice. Exp Neurol 203:333–348

Blinzinger K, Kreutzberg G (1968) Displacement of synaptic terminals from regenerating motoneurons by microglial cells. Z Zellforsch Mikrosk Anat 85:145–157

Bontioti EN, Kanje M, Dahlin LB (2003) Regeneration and functional recovery in the upper extremity of rats after various types of nerve injuries. J Peripher Nerv Syst 8:159–68

Bontioti E, Kanje M, Lundborg G, Dahlin LB (2005) End-to-side nerve repair in the upper extremity of rat. J Peripher Nerv Syst 10:58–68

Borgens RB (1988) Voltage gradients and ionic currents in injured and regenerating axons. Adv Neurol 47:51–66

Botti G (2006) Superior blepharoplasty. In: Vuyk HD, Lohuis PJFM (eds) Facial plastic and reconstructive surgery. Edward Arnold Publishers, London, pp 57–67

Bowe CM, Kocsis JD, Waxman SG (1985) Differences between mammalian ventral and dorsal spinal roots in response to blockade of potassium channels during maturation. Proc R Soc Lond 224:355–366

Brännström T, Kellerth JO (1998) Changes in synaptology of adult cat spinal alpha-motoneurons after axotomy. Exp Brain Res 118:1–13

Brännström T, Kellerth JO (1999) Recovery of synapses in axotomized adult cat spinal motoneurons after reinnervation into muscle. Exp Brain Res 125:19–27

Brännström T, Havton L, Kellerth JO (1992a) Changes in size and dendritic arborization patterns of adult cat spinal alpha-motoneurons following permanent axotomy. J Comp Neurol 318:439–451

Brännström T, Havton L, Kellerth JO (1992b) Restorative effects of reinnervation on the size and dendritic arborization patterns of axotomized cat spinal alpha-motoneurons. J Comp Neurol 318:452–461

Bratzlavsky M, vander Eecken H (1977) Altered synaptic organization in facial nucleus following facial nerve regeneration: an electrophysiological study in man. Ann Neurol 2:71–73

Bray GM, Aguayo AJ (1974) Regeneration of peripheral unmyelinated nerves. Fate of the axonal sprouts which develop after surgery. J Anat 117:517–529

Bray D, Bunge MB (1981) Serial analysis of microtubules of cultured rat sensory neurons. J Neurocytol 10:589–605

Brecht M, Preilowski B, Merzenich MM (1997) Functional architecture of the mystacial vibrissae. Behav Brain Res 84:81–97

Brennan RJ, Shirley JP, Compton JS (1993) Bilateral hypoglossal nerve palsy following head injury. J Emerg Med 11:167–168

Brown MC, Holland RL (1979) A central role for denervated tissues in causing nerve sprouting. Nature 282:724–726

Brown MC, Ironton R (1977) Motor neurone sprouting induced by prolonged tetrodotoxin block of nerve action potentials. Nature 265:459–461

Brown MC, Goodwin GM, Ironton R (1977) Prevention of motor nerve sprouting in botulinum toxin poisoned mouse soleus muscles by direct stimulation of the muscle. J Physiol (Lond) 267:42P–43P

Brown MC, Holland RL, Ironton R (1980) Nodal and terminal sprouting from motor nerves. J Physiol 306:493–510

Brown MC, Holland RL, Hopkins WG, Keynes RJ (1981) An assessment of the spread of the signal for terminal sprouting within and between muscles. Brain Res 210:145–151

Brushart TM, Mesulam MM (1980) Alteration in connections between muscle and anterior horn motoneurons after peripheral nerve repair. Science 208:603–605

Brushart TM, Seiler WA (1987) Selective reinnervation of distal motor stumps by peripheral motor axons. Exp Neurol 97:289–300

Brushart TM, Gerber J, Kessens P, Chen Y-G, Royall RM (1998) Contributions of pathway and neuron to preferential motor reinnervation. J Neurosci 18:8674–8681

Brushart TM, Hoffman PN, Royall RM, Murinson BB, Witzel C, Gordon T (2002) Electrical stimulation promotes motoneuron regeneration without increasing its speed or conditioning the neuron. J Neurosci 22:6631–6638

Brushart TM, Jari R, Verge V, Rohde C, Gordon T (2005) Electrical stimulation restores the specificity of sensory axon regeneration. Exp Neurol 194:221–229

Burton PR, Hinkley RE, Pierson GB (1975) Tannic acid stained microtubules with 12, 13, and 15 protofilaments. J Cell Biol 65:227–233

Calhoun ME, Jucker M, Martin LJ, Thinakaran G, Price DL, Mouton PR (1996) Comparative evaluation of synaptophysin-based methods for quantification of synapses. J Neurocytol 25:821–828

Cangiano A, Magherini PC, Pasino E, Pellegrino M, Risaliti R (1984) Interaction of inactivity and nerve breakdown products in the origin of acute denervation changes in rat skeletal muscle. J Physiol 355:3435–365

Carvell GE, Simons DJ (1990) Biometric analyses of vibrissal tactile discrimination in the rat. J Neurosci 10:2638–2648

Castro-Lopes JM, Tavares I, Coimbra A (1993) GABA decreases in the spinal cord dorsal horn after peripheral neurectomy. Brain Res 620:287–291

Challacombe JF, Snow DM, Letourneau PC (1996) Actin filament bundles are required for microtubule reorientation during growth cone turning to avoid an inhibitory guidance cue. J Cell Sci 109:2031–2040

Challacombe JF, Snow DM, Letourneau PC (1997) Dynamic microtubule ends are required for growth cone turning to avoid an inhitory guidance cue. J Neurosci 17:3085–3095

Choi D, Raisman G (2003) Immunre rejection of a facial nerve xenograft does not prevent regeneration and the return of function: an experimental study. Neuroscience 121:501–507

Choi D, Raisman G (2005) Disorganization of the facial nucleus after nerve lesioning and regeneration in the rat: effects of transplanting candidate reparative cells to the site of injury. Neurosurgery 56:1093–100

Chuckowree JA, Vickers JC (2003) Cytoskeletal and morphological alterations underlying axonal sprouting after localized transection of cortical neuron axons in vitro. J Neurosci 23:3715–25

Clatterbuck RE, Price DL, Koliatsos VE (1994) Further characterization of the effects of BDNF and CNTF on axotomized neonatal and adult mammalian motor neurons. J Comp Neurol 342:45–56

Cohan CS, Kater SB (1986) Suppression of neurite elongation and growth cone motility by electrical activity. Science 232:1638–1640

Cole BG, Gardiner PF (1984) Does electrical stimulation of denervated muscle, continued after reinnervation, influence recovery of contractile function? Exp Neurol 85:52–62

Cole J, Zimmerman S, Gerson S (1991) Nonsurgical neuromuscular rehabilitation of facial muscle paresis. In: Rubin LR (ed) The Paralyzed Face. Mosby-Year Book, St. Louis, MO, pp 107–112

Conley JJ, Baker D (1979) Hypoglossal-facial nerve anastomosis for reinnervation of the paralysed face. Plast Reconstr Surg 63:63–72

Connold AL, Vrbova G (1990) The effect of muscle activity on motor unit size in partially denervated rat soleus muscles. Neuroscience 34:525–532

Coulson SE (2005) Physiotherapy rehabilitation following facial nerve paresis. In: Beurskens CHG, van Gelder RS, Heymans PG, Manni J, Nicolai JPA (eds) The facial palsies. Complementary approaches. Lemma Publishers, Utrecht, The Netherlands, pp 263–274

Cunningham EJ, Bond R, Mayberg MR, Warlow CP, Rothwell PM (2004) Risk of persistent cranial nerve injury after carotid endarterectomy. J Neurosurg 101:445–448

D'Cruz PM, Yasumura D, Weir J, Matthes MT, Abderrahim H, La Vail MM, Vollrath D (2000) Mutation of the receptor tyrosine kinase gene Mertk in the retinal dystrophic RCS rat. Hum Mol Genet 9:645–651

Dauvergne C, Pinganaud G, Buisseret P, Buisseret-Delmas C, Zerari-Mailly F (2001) Reticular premotor neurons projecting to both facial and hypoglossal nuclei receive trigeminal afferents in rats. Neurosci Lett 311:109–112

de Bilbao F, Dubois-Dauphin M (1996) Time course of axotomy-induced apoptotic cell death in facial motoneurons of neonatal wild type and bcl-2 transgenic mice. Neuroscience 71:1111–1119

Decker MM, Berman HA (1990) Denervation-induced alterations of acetylcholinesterase in denrvated and nondenervated muscle. Exp Neurol 109:247–255

Deckwerth TL, Elliott JL, Knudson CM, Johnson EM Jr, Snider WD, Korsmeyer SJ (1996) BAX is required for neuronal death after trophic factor deprivation and during development. Neuron 17:401–411

Dedkov EI, Kostrominova TY, Borisov AB, Carlson BM (2001) Reparative myogenesis in long-term denervated skeletal muscles of adult rats results in a reduction of the satellite cell population. Anat Rec 263:139–154

Dent EW, Kalil K (2001) Axon branching requires interactions between dynamic microtubules and actin filaments. J Neurosci 21:9757–9769

Dent EW, Callaway JL, Szebenyi G, Baas PW, Kalil K (1999) Reorganization and movement of microtubules in axonal growth cones and developing interstitial branches. J Neurosci 19:8894–8908

Deschenes MR, Tenny KA, Wilson MH (2006) Increased and decreased activity elicits specific morphological adaptations of the neuromuscular junction. Neuroscience 137:1277–1283

Devor M, Keller CH, Deerinck TJ, Ellisman MH (1989) Na^+ channel accumulation on axolemma of afferent endings in nerve end neuromas in Apteronotus. Neurosci Lett 102:149–154

Diamond J, Coughlin M, Macintyre L, Holmes M, Visheau B (1987) Evidence that endogenous B nerve growth factor is responsible for the collateral sprouting, but not for regeneration of nociceptive axons in adult rats. Proc Natl Acad Sci USA 84:6596–6600

Diamond J, Holmes M, Coughlin M (1992) Endogenous NGF and nerve impulses regulate the collateral sprouting of sensory axons in the skin of the adult rat. J Neurosci 12:1454–1466

Dickens P, Hill P, Bennett MR (2003) Schwann cell dynamics with respect to newly formed motor-nerve terminal branches on mature (Bufo marinus) muscle fibers. J Neurocytol 32:381–392

Diels HJ (1995) New concepts in nonsurgical facial nerve rehabilitation. In: Myers E, Bluestone C (eds) Advances in otolaryngology-head and neck surgery. Mosby-Year Book, Chicago, IL, pp 289–313

Diels HJ (2005) Current concepts in non-surgical facial nerve rehabilitation. In: Beurskens CHG, van Gelder RS, Heymans PG, Manni JJ, Nicolai JPA (eds) The facial palsies. Complementary approaches. Lemma Publishers, Utrecht, pp 275–283

Dillon D, Gibbs MA, Baby C (2007) Peripheral nerve blocks of the hand. Acad Emerg Med 14:14–15

Ding M, Hart RP, Shadiack AM, Jonakait GM (1994) The interleukin-1-induced increase of substance P in sympathetic ganglia is not mediated by ciliary neurotrophic factor. J Neurosci Res 38:640–647

Dohm S, Streppel M, Guntinas-Lichius O, Pesheva P, Probstmeier R, Neiss WF, Angelov DN (2000) Local application of extracellular matrix proteins fails to reduce the number of axonal branches during regeneration. Restor Neurol Neurosci 16:117–126

Donoghue JP, Wise SP (1982) Rat motor cortex: cytoarchitecture and microstimulation mapping. J Comp Neurol 212:76–88

Dörfl J (1982) The musculature of the mystacial vibrissae of the white mouse. J Anat 135:147–154

Dörfl J (1985) The innervation of the mystacial region of the white mouse. A topographical study. J Anat 142:173–184

Dow DE, Carlson BM, Hassett CA, Dennis RG, Faulkner JA (2006) Electrical stimulation of denervated muscles of rats maintains mass and force, but no recovery following grafting. Restor Neurol Neurosci 24:41–54

Dunlop SA, Steeves JD (2003) Neural activity and facilitated recovery following CNS injury: Implications for rehabilitation. In: Bresnahan JC (ed) Topics in spinal cord rehabilitation, vol 8. A Thomas Land, St Louis, MO, pp 92–103

Dyck PJ, Hopkins AP (1972) Electron microscopic observations on degeneration and regeneration of unmyelinated nerve fibers. Brain 95:223–234

Eberhardt KA, Irintchev A, Al-Majed AA, Simova O, Brushart TM, Gordon T, Schachner M (2006) BDNF/TrkB signaling regulates HNK-1 carbohydrate expression in regenerating motor nerves and promotes functional recovery after peripheral nerve repair. Exp Neurol 198:500–10

Eberstein A, Eberstein S (1996) Electrical stimulation of denervated muscle: is it worthwhile? Med Sci Sports Exerc 28:1463–1469

Eccles JC (1944) Investigations on muscle atrophies arising from disuse and tenotomy. J Physiol 103:253–266

Eccles JC, Libet B, Young RR (1958) The behaviour of chromatolyzed motoneurons studied by intracellular recording. J Physiol 143:11–40

Einsiedel LJ, Luff AR (1994) Activity and motor unit size in partially denervated rat medial gastrocnemius. J Appl Physiol 76:2663–2671

Engel AK, Kreutzberg GW (1986) Changes in acetylcholinesterase molecular forms in regenerating motor neurons. Neuroscience 18:467–473

Engel AK, Tetzlaff W, Kreutzberg GW (1988) Axonal transport of 16S acetylcholinesterase is increased in regenerating peripheral nerve in guinea-pig, but not in the rat. Neuroscience 24:729–738

English AW (2003) Cytokines, growth factors and sprouting at the neuromuscular junction. J Neurocytol 32:943–60

English AW (2005) Enhancing axon regeneration in peripheral nerves also increases functionally inappropriate reinnervation of targets. J Comp Neurol 490:427–441

English AW, Schwartz G, Meador W, Sabatier MJ, Mulligan A (2007) Electrical stimulation promotes peripheral axon regeneration by enhanced neuronal neurotrophin signaling. Dev Neurobiol 67:158–172

Erzurumlu RS, Killackey HP (1979) Efferent connections of the brainstem trigeminal complex with the facial nucleus of the rat. J Comp Neurol 188:75–86

Esslen E (1960) Electromyographic findings on two types of misdirection of regenerating axons. EEG Clin Neurophysiol 12:738–741

Evgenieva E, Schweigert P, Guntinas-Lichius O, Pavlov SP, Grosheva M, Angelova SK, Streppel M, Irintchev A, Skouras E, Sinis N, Dunlop SA, Radeva V, Angelov DN (2008) Manual stimulation of tongue muscles diminishes polynnervation of the motor endplates and improves recovery of function after hypoglossal nerve injury in rats. Neurorehab Neural Repair 22:754–68

Farragher D, Kidd GL, Tallis R (1987) Eutrophic electrical stimulation for Bell's Palsy. Clin Rehab 1:265–271

Fawcett JW, Keynes RJ (1990) Peripheral nerve regeneration. Annu Rev Neurosci 13:43–60

Fenrich K, Gordon T (2004) Canadian Association of Neuroscience review: axonal regeneration in the peripheral and central nervous systems – current issues and advances. Can J Neurol Sci 31:142–156

Ferreira MC, Besteiro JM, Tuma Junior P (1994) Results of reconstruction of the facial nerve. Microsurgery 15:5–8

Fisch U, Lanser MJ (1991) Facial nerve grafting. Otolaryngol Clin North Am 24:691–708

Forscher P, Smith SJ (1988) Actions of cytochalasins on the organization of actin filaments and microtubules in a neuronal growth cone. J Cell Biol 107:1505–1516

Frach JP, Osterbauer PJ, Fuhr AW (1992) Treatment of Bell's palsy by mechanical force, manually assisted chropractic adjusting and high-voltage electrotherapy. J Manipulative Physiol Ther 15:596–598

Franchi G (2000) Reorganization of vibrissal motor representation following severing and repair of the facial nerve in adult rats. Exp Brain Res 131:33–43

Friede RL, Bischhausen R (1980) The fine structure of stumps of transected nerve fibers in subserial sections. J Neurol Sci 44:181–203

Fu SY, Gordon T (1997) The cellular and molecular basis of peripheral nerve regeneration. Mol Neurobiol 14:67–116

Fukata Y, Itoh TJ, Kimura T, Menager C, Nishimura T, Shiromizu T, Watanabe H, Inagaki N, Iwamatsu A, Hotani H, Kaibuchi K (2002) CRMP-2 binds to tubulin heterodimers to promote microtubule assembly. Nature Cell Biol 4:583–591

Gallo G, Letourneau PC (1998) GTPases help axons reach their targets. Curr Biol 8:R80–R82

Gallo G, Letourneau PC (1999) Different contributions of microtubule dynamics and transport to the growth of axons and collateral sprouts. J Neurosci 19:3860–3873

Galtrey CM, Asher RA, Nothias F, Fawcett JW (2007) Promoting plasticity in the spinal cord with chondroitinase improves functional recovery after peripheral nerve repair. Brain 130:926–39

Gao P, Bermejo R, Zeigler HP (2001) Whisker deafferentation and rodent whisking patterns: behavioral evidence for a central pattern generator. J Neurosci 21:5374–5380

Geremia NM, Gordon T, Brushart TM, Al-Majed AA, Verge VM (2007) Electrical stimulation promotes sensory neuron regeneration and growth-associated gene expression. Exp Neurol 205:347–59

Gittins J, Martin K, Sheldrick J, Reddy A, Thean L (1999) Electrical stimulation as a therapeutic option to improve eyelid function in chronic facial nerve disorders. Invest Ophthalmol Vis Sci 40:547–54

Goldberg DJ, Burmeister DW (1986) Stages in axon formation: observation of growth cones of Aplysia axons in culture using video-enhanced contrast-differential interference contrast microscopy. J Cell Biol 104:1921–1931

Gong S, Zhou Q, LeDoux MS (2003) Blink-related sensorimotor anatomy in the rat. Anat Embryol 207:193–208

Goodmurphy GW, Ovalle WK (1999) Morphological study of two human facial muscles: orbicularis oculi and corrugator supercilii. Clin Anat 12:1–11

Gordon T, Hegedus J, Tam SL (2004) Adaptive and maladaptive motor axonal sprouting in aging and motoneuron disease. Neurol Res 26:174–85

Gordon T, Brushart TM, Amirjani N, Chan KM (2007) The potential of electrical stimulation to promote functional recovery after peripheral nerve injury – comparisons between rats and humans. Acta Neurochir Suppl 100:3–11

Gordon T, Brushart TM, Chan KM (2008) Augmenting nerve regeneration with electrical stimulation. Neurol Res 30:1012–1020

Gordon-Weeks PR (1991) Evidence for microtubule capture by filopodial actin filaments in growth cones. NeuroReport 2:573–576

Gordon-Weeks PR (1997) Growth cones – the mechanism of neurite advance. Bioessays 13:235–239

Gorio A, Carmignoto G, Finesso M, Polato P, Nunzi MG (1983) Muscle reinnervation – II. Sprouting, synapse formation and repression. Neuroscience 8:403–416

Graeber MB, Kreutzberg GW (1988) Delayed astrocyte reaction following facial nerve axotomy. J Neurocytol 17:209–20

Graeber MB, Bise K, Mehraein P (1993) Synaptic stripping in the human facial nucleus. Acta Neuropathol (Berl) 86:179–181

Grant GA, Rostomily RR, Kim DK, Mayberg MR, Farrell D, Avellino A, Duckert LG, Gates GA, Winn HR (2002) Delayed facial palsy after resection of vestibular schwannoma. J Neurosurg 97:93–96

Greene EC (1935) Anatomy of the rat. Transactions of the American Philosophical Society, Philadelphia, New Series, Volume XXVII

Greene EC (1955) Anatomy of the Rat. New York: Hafner Publishing Co. Inc.

Griffin JW, Thompson WJ (2008) Biology and pathology of nonmyelinating Schwann cells. Glia 56:1518–1531

Grimby G, Einarsson G, Hedberg M, Aniansson A (1989) Muscle adaptive changes in post-polio subjects. Scand J Rehabil Med 21:19–26

Grosheva M, Arnhold S, Guntinas-Lichius O, Irintchev A, Skouras E, Kuerten S, Streppel M, Angelova SK, Wewetzer K, Radtke C, Dunlop SA, Angelov DN (2008) Transplantation of bone marrow-derived mesenchymal stem cells (BM-MSC) does not improve quality of muscle reinnervation and recovery of motor function after transection of the facial nerve in rats. Biol Chem 389:873–888

Gruart A, Streppel M, Guntinas-Lichius O, Angelov DN, Neiss WF, Delgado-Garcia JM (2003) Motoneuron adaptability to new motor tasks following two types of facial-facial anastomosis cats. Brain 126:115–133

Guan KL, Rao Yi (2003) Signalling mechanisms mediating neuronal responses to guidance cues. Nat Rev Neurosci 4:941–956

Guidry C, Grinnell F (1987) Heparin modulates the organization of hydrated collagen gels and inhibits gel contraction by fibroblasts. J Cell Biol 104:1097–1103

Gundersen HJ (1986) Stereology of arbitrary particles. A review of unbiased number and size estimators and the presentation of some new ones, in memory of William R. Thompson. J Microsc 143:3–45

Gundersen HJG, Bendtsen TF, Korbo L, Marcussen N, Moller A, Nielsen K, Nyengaard JR, Pakkenberg B, Sorensen FB, Vesterby A, West MJ (1988) Some new, simple and efficient stereological methods and their use in pathological research and diagnosis. Acta Pathol Microbiol Immunol Scand 96:379–394

Guntinas-Lichius O (2004) The facial nerve in the presence of a head and neck neoplasm: assessment and outcome after surgical treatment. Curr Opin Otolaryngol Head Neck Surg 12:133–141

Guntinas-Lichius O, Hundeshagen G, Paling T, Streppel M, Grosheva M, Irintchev AP, Skouras A, Alvanou A, Angelova SK, Kuerten S, Sinis N, Dunlop SA, Angelov DN (2007) Manual stimulation of facial muscles improves functional recovery after hypoglossal-facial anastomosis and interpositional nerve grafting of the facial nerve in adult rats. Neurobiol Disease 28:101–112

Guntinas-Lichius O, Irintchev A, Streppel M, Lenzen M, Grosheva M, Wewetzer K, Neiss WF, Angelov DN (2005) Factors limiting motor recovery after facial nerve transection in the rat: combined structural and functional analyses. Eur J Neurosci 21:391–402

Guntinas-Lichius O, Mockenhaupt J, Stennert E, Neiss WF (1993) Simplified nerve cell counting in the rat brainstem with the physical dissector using a drawing-microscope. J Microsc 172:177–180

Guntinas-Lichius O, Neiss WF, Gunkel A, Stennert E (1994) Differences in glial, synaptic and motoneuron responses in the facial nucleus of the rat brainstem following facial nerve resection and nerve suture reanastomosis. Eur Arch Otorhinolaryngol 251:410–417

Guntinas-Lichius O, Wewetzer K, Tomov TL, Azzolin A, Kazemi S, Streppel M, Neiss WF, Angelov DN (2002) Transplantation of olfactory mucosa minimizes axonal branching and promotes the recovery of vibrissae motor performance after facial nerve repair in rats. J Neurosci 22:7121–7131

Hahn CM, Kleinholz H, Koester MP, Grieser S, Thelen K, Pollerberg GE (2005) Role of cyclin-dependent kinase 5 and its activator P35 in local axon and growth cone stabilization. Neuroscience 134:449–465

Halloran MC, Kalil K (1994) Dynamic behaviors of growth cones extending in the corpus callosum of living cortical brain slices observed with video microscopy. J Neurosci 14:2161–2177

Hattox AM, Priest CA, Keller A (2002) Functional circuitry involved in the regulation of whisker movements. J Comp Neurol 442:266–76

Heidemann SR, Landers JM, Hamborg MA (1981) Polarity orientation of axonal microtubules. J Cell Biol 91:661–665

Hennig R (1987) Late reinnervation of the rat soleus muscle is differentially suppressed by chronic stimulation and by ectopic innervation. Acta Physiol Scand 130:153–60

Hennig R, Dietrichs E (1994) Transient reinnervation of antagonistic muscles by the same motoneuron. Exp Neurol 130:331–336

Hennig R, Lomo T (1985) Firing patterns of motor units in normal rats. Nature 314:164–166

Herbison GJ, Teng CS, Gordon EE (1973) Electrical stimulation of reinnervating rat muscle. Arch Phys Med Rehabil 54:156–160

Hinrichsen CF, Watson D (1983) Brain stem projections to the facial nucleus of the rat. Brain Behav Evol 22:153–163

Hinrichsen CF, Watson CD (1984) The facial nucleus of the rat: representation of facial muscles revealed by retrograde transport of horseradish peroxidase. Anat Rec 209:407–415

Hirata K, Kawabuchi M (2002) Myelin phagocytosis by macrophages and non-macrophages during Wallerian degeneration. Microsc Res Tech 57:541–547

Hirokawa N, Takemura R (2005) Molecular motors and mechanisms of directional transport in axons. Nat Rev Neurosci 6:201–214

Hoffman PN, Cleveland DW (1988) Neurofilament and tubulin gene expression recapitulates the developmental program during axonal regeneration. Induction of a specific β-tubulin isotype. Proc Natl Acad Sci USA 85:4530–4533

Hoffman PN, Lasek RJ (1980) Axonal transport of the cytoskeleton in regenerating motor neurons. Constancy and change. Brain Res 202:317–333

Hoffman PN, Lopata MA, Watson DE, Luduena RF (1992) Axonal transport of Class II and III β-tubulin: evidence that the slow component wave represents the movement of only a small fraction of the tubulin in mature motor axons. J Cell Biol 119:595–604

Horch KWL, Lisney SJW (1981) On the number and nature of regenerating unmyelinated axons after lesions of the cutaneous nerves in the cat. J Physiol 313:275–286

Horvath S, Prandovszky E, Pankotai E, Kis Z, Farkas T, Boldoköi Z, Boda K, Janka Z, Toldi J (2005) Use of recombinant pseudorabies virus to analyze motor cortical reorganization after unilateral facial denervation. Cerebr Cortex 15:378–384

Hosemann W, Wigand ME, Herrlinger P, Weckopp M (1990) Surgical reneurotization of the tongue after unilateral paralysis of the hypoglossal nerve. Eur Arch Otorhinolaryngol 248:95–98

Hovind H, Nielsen S (1974) Effect of massage on blood flow in skeletal muscle. Scand J Rehab Med 6:74–77

Huizing EH, Mechelse K, Staal A (1981) Treatment of Bell's Palsy. An analysis of the available studies. Acta Otolaryngol 92:115–121

Ijkema-Paassen J, Meek MF, Gramsbergen A (2002) Reinnervation of muscles after transection of the sciatic nerve in adult rats. Muscle Nerve 25:891–897

Illert M, Trauner M, Weller E, Wiedemann E (1986) Forearm muscles of man can reverse their function after tendon transfers: an electromyographic study. Neurosci Lett 67:129–134

Irintchev A, Draguhn A, Wernig A (1990) Reinnervation and recovery of mouse soleusmuscle after long-term denervation. Neuroscience 39:231–243

Ishikawa M, Namiki J, Takase M, Ohira T, Nakamura A, Toya S (1996) Effect of repetitive stimulation on lateral spread of F-waves in hemifacial spasm. J Neurol Sci 142:99–106

Isokawa-Akesson M, Komisaruk BR (1987) Difference in projections to the lateral and medial facial nucleus: anatomically separate pathways for rhythmical vibrissa movement in rats. Exp Brain Res 65:385–398

Ito M, Kudo M (1994) Reinnervation by axon collaterals from single facial motoneurons to multiple muscle targets following axotomy in the adult guinea pig. Acta Anat (Basel) 151:124–130

Jacquin MF, Zahm DS, Henderson TA, Golden JP, Johnson EM, Renehan WE, Klein BG (1993) Structure-function relationships in rat brainstem subnucleus interpolaris. X. Mechanisms underlying enlarged spared whisker projections after infraorbital nerve injury at birth. J Neurosci 13:2946–2964

Jenq C-B, Jenq LL, Bear HM, Coggeshall RE (1988) Conditioning lesions of peripheral nerves change regenerated axon numbers. Brain Res 457:63–69

Jergovic D, Stal P, Lidman D, Lindvall B, Hildebrand C (2001) Changes in a rat facial muscle after facial nerve injury and repair. Muscle Nerve 24:1202–1212

Jevsek M, Mars T, Mis K, Grubic Z (2004) Origin of acetylcholinesterase in the neuromuscular junction formed in the in vitro innervated human muscle. Eur J Neurosci 20:2865–2871

Jin Z, Strittmatter SM (1997) Rac 1 mediates collapsin-1 induced growth cones collapse. J Neurosci 17:6256–6263

Kalil K, Dent EW (2005) Touch and go: guidance cues signal to the growth cone cytoskeleton. Curr Opin Neurobiol 15:521–526

Kang H, Tian L, Thompson W (2003) Terminal Schwann cells guide the reinnervation of muscle after nerve injury. J Neurocytol 32:975–985

Keane JR (1996) Twelfth-nerve palsy. Analysis of 100 cases. Arch Neurol 53:561–566

Kelly EJ, Jacoby C, Terenghi G, Mennen U, Ljungberg C, Wiberg M (2007) End-to-side nerve coaptation: a qualitative and quantitative assessment in the primate. J Plast Reconstr Aesthet Surg 60:1–12

Kern H, Hofer C, Modlin M, Forstner C, Raschka-Hogler D, Mayr W, Stohr H (2002) Denervated muscles in humans: limitations and problems of currently used functional electrical stimulation training protocols. Artif Organs 26:216–218

Kern H, Salmons S, Mayr W, Rossini K, Carraro U (2005) Recovery of long-term denervated human muscles induced by electrical stimulation. Muscle Nerve 31:98–101

Kerrebijn JDF, Freeman JL (1998) Facial nerve reconstruction: outcome and features. J Otolaryngol 27:183–186

Kimura J, Lyon LW (1972) Orbicularis oculi reflex in the Wallenberg syndrome: alteration of the late reflex by lesions of the spinal tract and nucleus of the trigeminal nerve. J Neurol Neurosurg Psychiatry 35:228–233

Kimura J, Rodnitzky RL, Okawara SH (1975) Electrophysiologic analysis of aberrant regeneration after facial nerve paralysis. Neurology 25:989–993

King CE, Canty AJ, Vickers JC (2001) Alterations in neurofilament associated with reactive brain changes and axonal sprouting following acute physical injury to the rat neocortex. Neuropathol Appl Neurobiol 27:115–126

Kis Z, Rkos G, Farkas T, Horvth S, Toldi J (2004) Facial nerve injury induces facilitation of responses in both trigeminal and facial nuclei of rat. Neurosci Lett 358:223–225

Kitzman P (2005) Alteration in axial motoneuron morphology in the spinal cord of the injured spastic rat. Exp Neurol 192:100–108

Klein BG, Rhoades RW (1985) Representation of whisker follicle intrinsic musculature in the facial motor nucleus of the rat. J Comp Neurol 232:55–69

Kleinfeld D, Berg RW, O'Connor SM (1999) Anatomical loops and their electrical dynamics in relation to whisking by rat. Somatosens Mot Res 16:69–88

Komisaruk BR (1970) Synchrony between limbic system theta activity and rhythmical behaviour in rats. J Comp Physiol Psychol 70:482–492

Krammer EB, Rath T, Lischka MF (1979) Somatotopic organization of the hypoglossal nucleus: a HRP study in the rat. Brain Res 170:533–537

Krarup C, Archibald SJ, Madison RD (2002) Factors that influence peripheral nerve regeneration: an electrophysiological study of the monkey median nerve. Ann Neurol 51:69–81
Kreutzberg GW, Toth L, Kaiya H (1975) Acetylcholinesterase as a marker for dendritic transport and dendritic secretion. Adv Neurol 12:269–281
Kunihiro T, Higashino K, Kanzaki J (2003) Classic hypoglossal-facial nerve anastomosis after neurona resection. A review of 46 cases. ORL J Otolaryngol Relat Spec 65:1–6
Kuno M, Llinas R (1970) Enhancement of synaptic transmission by dendritic potentials in chromatolysed motoneurones of the cat. J Physiol (Lond) 210:807–821
Kuroki A, Moller AR, Saito S (1994) Recordings from the facial motonucleus in rats with signs of hemifacial spasm. Neurol Res 16:389–392
Lavy J, East CA, Bamber A, Andrews PJ (2004) Gold weight implants in the management of lagophtalmus in facial palsy. Clin Otolaryngol 29:279–283
Leiser SC, Moxon KA (2007) Responses of trigeminal ganglion neurons during natural whisking behaviors in the awake rat. Neuron 55:117–133
Letourneau PC, Cypher C (1991) Regulation of growth cone motility. Cell Motil Cytoskeleton 20:267–271
Li Y, Decherchi P, Raisman G (2003) Transplantation of olfactory ensheathing cells into spinal cord lesions restores breathing and climbing. J Neurosci 23:727–731
Lieber RL (1992) Skeletal muscle structure and function: implications for rehabilitation and sports medicine. Williams and Wilkins, Baltimore, MD, pp 210–259
Lin CH, Forscher P (1993) Cytoskeletal remodeling during growth cone-target interactions. J Cell Biol 121:1369–1383
Lin CH, Thompson CA, Forscher P (1994) Cytoskeletal reorganization underlying growth cone motility. Curr Opin Neurobiol 4:640–647
Lindsay FW, Mullin D, Keefe MA (2003) Subacute hypoglossal nerve paresis with internal carotid artery dissection. Laryngoscope 113:1530–1533
Lomo T, Slater CR (1980) Control of junctional acetylcholinesterase by neural and muscular influences in the rat. J Physiol 303:191–202
Lömo T, Westgaard RH (1974) Contractile properties of muscle: control by pattern of muscle activity in the rat. Proc R Soc Lond B187:99–103
Love FM, Thompson WJ (1999) Glial cells promote muscle reinnervation by responding to activity-dependent postsynaptic signals. J Neurosci 19:10390–10396
Love FM, Son YJ, Thompson WJ (2003) Activity alters muscle reinnervation and terminal sprouting by reducing the number of Schwann cell pathways that grow to link synaptic sites. J Neurobiol 54:566–576
Lowe AA (1981) The neural regulation of tongue movements. Progr Neurobiol 15:295–344
Lowe BD, Freivalds A (1999) Effect of carpal tunnel syndrome on grip force coordination on hand tools. Ergonomics 42:550–64
Lundborg G (2003) Richard P. Bunge memorial lecture. Nerve injury and repair – a challenge to the plastic brain. J Peripher Nerv Syst 8:209–226
Lux HD, Schubert P (1975) Some aspects of the electroanatomy of dendrites. Adv Neurol 12:29–44
Mackinnon SE, Dellon AL, Hudson AR, Hunter DA (1985) A primate model for chronic nerve compression. J Reconstr Microsurg 1:185–195
Mackinnon SE, Dellon AL, Obrien JP (1991) Changes in nerve fiber diameters distal to a nerve repair in the rat sciatic nerve model. Muscle Nerve 14:1116–1122
Mader K, Andermahr J, Angelov DN, Neiss WF (2004) Dual mode of signalling of the axotomy reaction: retrograde electric stimulation or block of retrograde transport differently mimic the reaction of motoneurons to nerve transection in the rat brainstem. J Neurotrauma 21:956–68
Madison RD, Archibald SJ, Lacin R, Krarup C (1999) Factors contributing to preferential motor reinnervation in the primate peripheral nervous system. J Neurosci 19:11007–11016

Manni JJ, Beuerskens CHG, van de Velde C, Stokroos RJ (2001) Reanimation of the paralyzed face by indirect hypoglossal-facial nerve anastomosis. Am J Surgery 182:268–273

Marques KB, Santos LMB, Oliveira ALR (2006) Spinal motoneuron synaptic plasticity during the course of an animal model of mutiple sclerosis. Eur J Neurosci 24:3053–3062

Marqueste T, Decherchi P, Dousset E, Berthelin F, Jammes Y (2002) Effect of muscle stimulation on afferent activities from tibialis anterior muscle after nerve repair by self-anastomosis. Neuroscience 113:257–271

Marqueste T, Decherchi P, Desplanches D, Favier R, Grelot L, Jammes Y (2006) Chronis electrostimulation after nerve repair by self-anastomosis: effects on the size, the mechanical, histochemical and biochemical muscle properties. Acta Neuropathol 111:589–600

Marsh L, Letourneau PC (1984) Growth of neurites without filopodial or lamellipodial activity in the presence of cytochalasin B. J Cell Biol 99:2041–2047

Masliah E, Terry RD, Alford M, DeTeresa R (1990) Quantitative immunohistochemistry of synaptophysin in human neocortex: an alternative method to estimate density of presynaptic terminals in paraffin sections. J Histochem Cytochem 38:837–44

Mauch C, Hatamoch A, Scharffetter K, Krieg T (1988) Regulation of collagen synthesis in fibroblasts within a three-dimensional collagen gel. Exp Cell Res 178:493–503

May M (1986) Surgical rehabilitation of facial palsy. In: May M (ed) The facial nerve. Thieme, New York, pp 695–777

McComas AJ (1998) Oro-facial muscles: internal structure, function and ageing. Gerodontology 15:3–14

McCulloch KL, Nelson CM (1995) Electrical stimulation and electromyographic biofeedback. In: Umphred DA (ed) Neurological Rehabilitation. Mosby Year Book, St. Louis, MO, pp 853–856

McHale MK, Hall GF, Cohen MJ (1995) Early cytoskeletal changes following injury of giant spinal axons in the lamprey. J Comp Neurol 353:25–37

McPhail LT, Stirling DP, Tetzlaff W, Kwiecien JM, Ramer MS (2004a) The contribution of activated phagocytes and myelin degeneration to axonal retraction/dieback following spinal cord injury. Eur J Neurosci 20:1984–1994

McPhail LT, Fernandes KJ, Chan CC, Vanderluit JL, Tetzlaff W (2004b) Axonal reinjury reveals the survival and re-expression of regeneration-associated genes in chronically axotomized adult mouse motoneurons. Exp Neurol 188:331–340

Mendonca AC, Barbieri CH, Mazzer N (2003) Directly applied low intensity direct electric current enhances peripheral nerve regeneration in rats. J Neurosci Methods 129:183–190

Merkler D, Metz GA, Raineteau O, Dietz V, Schwab M, Fouad K (2001) Locomotor recovery in spinal cord-injured rats treated with an antibody neutralizing the myelin-associated neurite growth inhibitor Nogo-A. J Neurosci 21:3665–3673

Miller AJ (2002) Oral and pharyngeal reflexes in the mammalian nervous system: their diverse range in complexity and the pivotal role of the tongue. Crit Rev Oral Biol Med 13:409–425

Minnery BS, Simons DJ (2003) Response properties of whisker-associated trigeminothalamic neurons in rat nucleus principalis. J Neurophysiol 89:40–56

Mokrusch T, Engelhardt A, Eichhorn K-F, Prischenk G, Sack G, Meindorfer B (1990) Effects of long-impulse electrical stimulation on atrophy and fibre type composition of chronically denervated fast rabbit muscle. J Neurol 237:29–34

Moller AR, Jannetta PJ (1986) Blink reflex in patients with hemifacial spasm. Observations during microvascular decompression operations. J Neurol Sci 72:171–182

Moller AR, Sen CN (1990) Recordings from the facial nucleus in the rat. Signs of abnormal facial muscle response. Exp Brain Res 81:18–24

Montserrat L, Benito M (1988) Facial synkinesis and aberrant regeneration of facial nerve. Adv Neurol 49:211–224

Moran LB, Graeber MB (2004) The facial nerve axotomy model. Brain Res Brain Res Rev 44:154–178

Moran CJ, Neely JG (1996) Patterns of facial nerve synkinesis. Laryngoscope 106:1491–1496

Moran LB, Kosel S, Spitzer C, Schwaiger FW, Riess O, Kreutzberg GW, Graeber MB (2001) Expression of alpha-synuclein in non-apoptotic, slowly degenerating facial motoneurones. J Neurocytol 30:515–521

Morin D, Monteau R, Hilaire G (1992) Compared effects of serotonin on cervical and hypoglossal inspiratory activities: an in vitro study in the newborn rat. J Physiol 451:605–629

Mosforth J, Taverner D (1958) Physiotherapy for Bell's palsy. Br Med J 9:675–679

Moskowitz PF, Smith R, Pickett J, Frankfurter A, Oblinger MM (1993) Expression of the class III β-tubulin gene during axonal regeneration of rat dorsal root ganglion neurons. J Neurosci Res 34:129–134

Mourits MP, Vuyk HD (2006) Eylid reconstruction. In: Vuyk HD, Lohuis PJFM (eds) Facial plastic and reconstructive surgery. Edward Arnold, London, pp 437–453

Munger BL, Renehan WE (1989) Degeneration and regeneration of peripheral nerve in the rat trigeminal system: III. Abnormal sensory reinnervation of rat guard hairs following nerve transection and crush. J Comp Neurol 283:169–176

Murai N, Ueba T, Takahashi JA, Yang H-Q, Kikuchi H, Hiai H, Hatanaka M, Fukumoto M (1996) Apoptosis of human glioma cells in vitro and in vivo induced by neutralizing antibody against human basic fibroblast growth factor. J Neurosurg 85:1072–1077

Murphy EH, Brown J, Iannuzzelli P, Baker R (1990) Regeneration and soma size changes following axotomy of the trochlear nerve. J Comp Neurol 292:524–536

Nacimiento W, Podoll K, Graeber MB, Topper R, Mobius E, Ostermann H, Noth J, Kreutzberg GW (1992) Contralateral early blink reflex in patients with facial nerve palsy: indication for synaptic reorganization in the facial nucleus during regeneration. J Neurol Sci 109:148–155

Narita M, Ozaki S, Narita M, Ise Y, Yajima Y, Suzuki T (2003) Change in the expression of c-fos in the brain following sciatic nerve ligation. Neurosci Lett 352:231–233

Neafsey EJ, Bold EL, Haas G, Hurley-Gius KM, Quirk G, Sievert CF, Terreberry RR (1986) The organization of the rat motor cortex: a microstimulation mapping study. Brain Res Rev 11:77–96

Neiss WF, Guntinas-Lichius O, Angelov DN, Gunkel A, Stennert E (1992) The hypoglossal-facial anastomosis as model of neuronal plasticity in the rat. Ann Anat 174:419–433

Nelson PG, Fields RD, Yu C, Liu Y (1993) Synapse elimination from the mouse neuromuscular junction in vitro: a non-Hebbian activity-dependent process. J Neurobiol 24:1517–30

Nguyen QT, Kleinfeld D (2005) Positive feedback in a brainstem tactile sensorimotor loop. Neuron 45:447–57

Nguyen QT, Sanes JR, Lichtman JW (2002) Pre-existing pathways promote precise projection patterns. Nat Neurosci 5:861–867

Nicolaidis SC, Williams HB (2001) Muscle preservation using an implantable electrical system after nerve injury and repair. Microsurgery 21:241–247

Nix WA (1990) Effects of intermittent high frequency electrical stimulation on denervated EDL muscle of rabbit. Muscle Nerve 13:580–585

Nowak DA, Hermsdorfer J (2003) Selective deficits of grip force control during object manipulation in patients with reduced sensibility of the grasping digits. Neurosci Res 47:65–72

O'Leary DDM, Terashima T (1988) Cortical axons branch to multiple subcortical targets by intestitial axon budding: implications for target recognition and "waiting periods". Neuron 1:901–910

O'Reilly PMR, Fitzgerald MJT (1990) Fibre composition of the hypoglossal nerve in the rat. J Anat 172:227–243

O'Malley JP, Waran MT, Balice-Gordon RJ (1999) *In vivo* observations of terminal Schwann cells at normal, denervated, and reinnervated mouse neuromuscular junctions. J Neurobiol 38:270–286

Oblinger MM, Lasek RJ (1988) Axotomy-induced alterations in the synthesis and transport of neurofilaments and microtubules in the dorsal root ganglion cells. J Neurosci 8:1746–1758

Pan YA, Misgeld T, Lichtman JW, Sanes JR (2003) Effects of neurotoxic and neuroprotective agents on peripheral nerve regeneration assayed by time-lapse imaging in vivo. J Neurosci 23:11479–11488

Papalia I, Tos P, Stagno d'Alcontres F, Battiston B, Geuna S (2003) On the use of the grasping test in the rat median nerve model: a re-appraisal of its efficacy for quantitative assessment of motor function recovery. J Neurosci Methods 127:43–47

Pavlov S, Grosheva M, Streppel M, Guntinas-Lichius O, Irintchev A, Skouras E, Angelova S, Kuerten S, Sinis N, Dunlop SA, Angelov DN (2008) Manually stimulated recovery of motor function after facial nerve injury requires intact sensory input. Exp Neurol 211:292–300

Peeva GP, Angelova SK, Guntinas-Lichius O, Streppel M, Irintchev A, Schütz U, Popratiloff A, Savaskan NE, Bräuer AU, Alvanou A, Nitsch R, Angelov DN (2006) Improved outcome of facial nerve repair in rats is associated with enhanced regenerative response of motoneurons and augmented neocortical plasticity. Eur J Neurosci 24:2152–62

Perry VH, Brown MC (1992) Role of macrophages in peripheral nerve degeneration and repair. Bioessays 14:401–406

Pockett S, Slack JR (1982) Source of the stimulus for nerve terminal sprouting in partially denervated muscle. Neuroscience 7:3173–3176

Popratiloff AS, Streppel M, Gruart A, Guntinas-Lichius O, Angelov DN, Stennert E, Delgado-Garcia JM, Neiss WF (2001) Hypoglossal and reticular interneurons involved in oro-facial coordination in the rat. J Comp Neurol 433:364–379

Raivich G, Makwana M (2007) The making of successful axonal regeneration: genes, molecules and signal transduction pathways. Brain Res Rev 53:287–311

Raivich G, Jones LL, Kloss CU, Werner A, Neumann H, Kreutzberg GW (1998) Immune surveillance in the injured nervous system: T-lymphocytes invade the axotomized mouse facial motor nucleus and aggregate around sites of neuronal degeneration. J Neurosci 18:5804–16

Raivich G, Bohatschek M, Kloss CU, Werner A, Jones LL, Kreutzberg GW (1999) Neuroglial activation repertoire in the injured brain: graded response, molecular mechanisms and cues to physiological function. Brain Res Brain Res Rev 30:77–105

Ramon-Cuéto A, Isabel Cordero M, Santos-Benito FF, Avila J (2000) Functional recovery of paraplegic rats and motor axon regeneration in their spinal cords by olfactory ensheathing glia. Neuron 25:425–435

Rathakrishnan R, Therimadasamy AK, Chan YH, Wilder-Smith EP (2007) The median palmar cutaneous nerve in normal subjects and CTS. Clin Neurophysiol 118:776–80

Reddy L, Koirala S, Siguira Y, Herrera AA, Ko C-P (2003) Glial cells maintain synaptic structure and function and promote development of the neuromuscular junction in vivo. Neuron 40:563–580

Reynolds ML, Woolf CJ (1992) Terminal Schwann cells elaborate extensive processes following denervation of the motor endplate. J Neurocytol 21:50–66

Rice FL, Kinnman E, Aldskogius H, Johansson O, Arvidsson J (1993) The innervation of the mystacial pad of the rat as revealed by PGP 9.5 immunofluorescence. J Comp Neurol 337:366–385

Rice FL, Fundin BT, Arvidsson J, Aldskogius H, Johansson O (1997) Comprehensive immunofluorescence and lectin binding analysis of vibrissal follicle sinus complex innervation in the mystacial pad of the rat. J Comp Neurol 385:149–184

Rich MM, Lichtman JW (1989) In vivo visualization of pre- and postsynaptic change during synapse elimination in reinnervated mouse muscle. J Neurosci 9:1781–1805

Ro L-S, Chen S-T, Tang L-M, Chang H-S (1996) Local application of anti-NGF blocks the collateral sprouting in rats following chronic construction injury of the sciatic nerve. Neurosci Lett 218:87–90

Rodel RM, Tergau F, Markus H, Laskawi R (2004) Bilateral changes in the cortical motor representation of the tongue after unilateral peripheral facial paralysis: evidence from transcranial magnetic stimulation. Ann Otol Rhinol Laryngol 113:951–955

Rossiter JP, Riopelle RJ, Bisby MA (1996) Axotomy-induced apoptotic cell death of neonatal rat facial motoneurons: time course analysis and relation to NADPH-diaphorase activity. Exp Neurol 138:33–44

Rotter A, Birdsall NJM, Burgen ASV, Field PM, Smolen A, Raisman G (1979) Muscarinic receptors in the central nervous system of the rat.IV. A comparison of the effects of axotomy and deafferentation on the binding of [^{3}H]propylbenzilylcholine mustard and associated synaptic changes in the hypoglossal and pontine nuclei. Brain Res Brain Res Rev 1:207–224

Rubio-Nazabal E, Marey-Lopez J, Lopez-Facal S, Alvarez-Perez P, Martinez-Figueroa A, Rey del Corral P (2002) Isolated bilateral paralysis of the hypoglossal nerve after transoral intubation for general anesthesia. Anesthesiology 96:245–247

Sadjadpour K (1975) Postfacial palsy phenomena: faulty nerve regeneration or ephaptic transmission? Brain Res 95:403–406

Sahly I, Khoutorsky A, Erez H, Prager-Khoutorsky M, Spira ME (2005) On-line confocal imaging of the events leading to structural dedifferentiation of an axonal segment into a growth cone after axotomy. J Comp Neurol 494:705–720

Saleh E, Achilli V, Naguib M (1995) Facial nerve neuromas: diagnosis and management. Am J Otol 16:521–6

Salmons S, Jarvis JC (2008) Functional electrical stimulation of denervated muscles: an experimental evaluation. Artif Organs 32:597–603

Salmons S, Ashley Z, Sutherland H, Russold MF, Li F, Jarvis JC (2005) Functional electrical stimulation of denervated muscles: basic issues. Artif Organs 29:199–202

Sanes JN, Donoghue JP (2000) Plasticity and primary motor cortex. Ann Rev Neurosci 23:393–415

Sanes JN, Suner S, Lando JF, Donoghue JP (1988) Rapid reorganization of adult rat motor cortex somatic representation patterns after motor nerve injury. Proc Natl Acad Sci USA 185:2003–2007

Sanes JN, Suner S, Donoghue JP (1990) Dynamic organization of primary motor cortex output to target muscles in adult rats. I. Long term patterns of reorganization following motor or mixed peripeheral nerve lesions. Exp Brain Res 79:479–91

Santos Da Silva J, Dotti CG (2002) Breaking the neuronal sphere: regulation of the actin cytoskeleton in neuritogenesis. Nat Rev Neurosci 3:694–704

Sawczuk A, Moser KM (2001) Neural control of tongue movement with respect to respiration and swallowing. Crit Rev Oral Biol Med 12:18–37

Schaefer AW, Kabir N, Forscher P (2002) Filopodia and actin arcs guide the assembly and transport of two populations of microtubules with unique dynamic parameters in neuronal growth cones. J Cell Biol 158:139–152

Schmalbruch H, al-Amood WS, Lewis DM (1991) Morphology of long-term denervated rat soleus muscle and the effect of chronic electrical stimulation. J Physiol 441:233–241

Schneider I, Gunkel A, Stennert E, Neiss WF (1994) Development of the nerve conduction velocity after hypoglossal-facial nerve anastomosis: an electro-neurographic study. Europ Arch Otorhinolaryngol Suppl p. 392.

Schröder JM (1968) Die Hyperneurotisation Büngnerscher Bänder bei der experimentallen Isoniazid Neuropathie: Phasenkontrast- und elektronenmikroskopische Untersuchungen. Virchows Arch Abt B1:131–156

Schwarting S, Schröder M, Stennert E, Goebel HH (1984) Morphology of denervated human facial muscles. Oto-Rhino-Laryngol 46:248–256

Schwarz JR, Bromm B, Spielmann P, Weytjens JL (1983) Development of Na^+ inactivation in motor and sensory myelinated nerve fibres of Rana esculenta. Pflugers Arch 398:126–129

Segev I (1998) Sound grounds for computing dendrites. Nature 393:207–208

Semba K, Egger MD (1986) The facial "motor" nerve of the rat: control of vibrissal movement and examination of motor and sensory components. J Comp Neurol 247:144–158

Semba K, Szechtman H, Komisaruk BR (1980) Synchrony among rhythmical facial tremor, neocortical "alpha" waves, and thalamic non-sensory neuronal bursts in intact awake rats. Brain Res 195:281–298

Senba E, Simmons DM, Wada E, Wada K, Swanson LW (1990) RNA levels of neuronal nicotinic acetylcholine receptor subunits are differentially regulated in axotomized facial motoneurons. An in situ hybridization study. Brain Res Mol Brain Res 8:349–353

Sendtner M (1998) Neurotrophic factors: Effects in modulating properties of the neuromuscular endplate. Cytokine Growth Factors Rev 9:1–7

Sengupta DK, Grevitt MP, Mehdian SM (1999) Hypoglossal nerve injury as a complication of anterior surgery to the upper cervical spine. Eur Spine 8:78–80

Sharp FR, Gonzalez MF, Morgan CW, Morton MT, Sharp JW (1988) Common fur and mystacial vibrissae parallel sensory pathways: 14 C 2-deoxyglucose and WGA-HRP studies in the rat. J Comp Neurol 270:446–469

Sharp CM, Borg HK, Kishore A, MacKenzie K (2002) Hypoglossal nerve paralysis following tonsillectomy. J Laryngol Otol 116:389–391

Shaw G, Bray D (1977) Movement and extension of isolated growth cones. Exp Cell Res 104:55–62

Shawe GD (1954) On the number of branches formed by regenerating nerve fibers. Br J Surg 42:474–488

Sheedlo HJ, Gaur V, Seaton AD, Turner JE (1991) Transplantation to the diseased and damaged retina. Trends Neurosci 14:347–350

Sherwood CC (2005) Comparative anatomy of the facial motor nucleus in mammals, with an analysis of neuron numbers in primates. Anat Rec 287A:1067–1079

Simova O, Irintchev A, Mehanna A, Liu J, Dihné M, Bächle D, Sewald N, Loers G, Schachner M (2006) Carbohydrate mimics promote functional recovery after peripheral nerve repair. Ann Neurol 60:430–437

Sinis N, Schaller HE, Schulte-Eversum C, Schlosshauer B, Doser M, Dietz K, Rosner H, Muller HW, Haerle M (2005) Nerve regeneration across a 2-cm gap in the rat median nerve using a resorbable nerve conduit filled with Schwann cells. J Neurosurg 103:1067–76

Sinis N, Guntinas-Lichius O, Irintchev A, Skouras E, Kuerten S, Pavlov SP, Schaller HE, Dunlop SA, Angelov DN (2008) Manual stimulation of forearm muscles does not improve recovery of motor function after injury to a mixed peripheral nerve. Exp Brain Res 185:469–483

Sinis N, Horn F, Genchev B, Skouras E, Angelova SK, Kaidoglou K, Michael J, Pavlov S, Igelmund P, Schaller HE, Kuerten S, Irintchev A, Dunlop SA, Angelov DN (2009) Electrical stimulation of paralyzed vibrissal muscles reduces endplate reinnervation and does not promote motor recovery after facial nerve repair in rats. Ann Anat 191:356–370

Sjoberg J, Kanje M (1990) The initial period of peripheral nerve regeneration and the importance of the local environment for the conditioning lesion effect. Brain Res 529:79–84

Skouras E, Popratiloff AS, Guntinas-Lichius O, Streppel M, Rehm KE, Neiss WF, Angelov DN (2002) Altered sensory input improves the accuracy of muscle reinnervation. Restor Neurol Neurosci 20:1–14

Skouras E, Merkel D, Grosheva M, Angelova SK, Schiffer G, Thelen U, Kaidoglou K, Sinis N, Igelmund P, Dunlop SA, Pavlov S, Irintchev A, Angelov DN (2009) Manual stimulation, but not acute electrical stimulation prior to reconstructive surgery, improves functional recovery after facial nerve injury in rats. Restor Neurol Neurosci 27:237–51

Slack JR, Pockett S (1981) Terminal sprouting of motoneurones is a local response to a local stimulus. Brain Res 217:368–374

Smith DS, Skene JHP (1997) A transcription dependent switch controls competence of adult neurons for distinct modes of axon growth. J Neurosci 17:646–658

Son YJ, Thompson WJ (1995) Nerve sprouting in muscle is induced and guided by processes extended by Schwann cells. Neuron 14:133–141

Son YJ, Trachtenberg JT, Thompson WJ (1996) Schwann cells induce and guide sprouting and reinnervation of neuromuscular junctions. Trends Neurosci 19:280–285

Sosnik R, Haidarliu S, Ahissar E (2001) Temporal frequency of whisker movement. I. Representations in brain stem and thalamus. J Neurophysiol 86:339–353

Spielmann RP, Schwarz JR, Bromm B (1983) Oscillating repolarization in action potentials of frog sensory myelinated fibres. Neurosci Lett 36:49–53

Spira ME, Oren R, Dormann A, Gitler D (2003) Critical calpain-dependent ultrastructural alterations underlie the transformation of an axonal segment into a growth cone after axotomy of cultured Aplysia neurons. J Comp Neurol 457:293–312

Spiwoks-Becker I, Vollrath L, Seeliger MW, Jaissle G, Eshkind LG, Leube RE (2001) Synaptic vesicle alterations in rod photoreceptors of synaptophysin-deficient mice. Neuroscience 107:127–142

Spruston N, Schiller Y, Stuart G, Sakmann B (1995) Activity-dependent action potential invasion and calcium influx into hippocampal CA1 dendrites. Science 268:297–300

Stal S, Peterson R, Spira M (1990) Aesthetic considerations and the pediatric population. Clin Plast Surg 17:133–149

Stal S, Spira M, Hamilton S (1987) Skin morphology and function. Clin Plast Surg 14:201–208

Standler NA, Bernstein JJ (1982) Degeneration and regeneration of motoneuron dendrites after ventral root crush: computer reconstruction of dendritic fields. Exp Neurol 75:600–615

Stennert E, Limberg CH (1982) Central connections between fifth, seventh, and twelfth cranial nerves and their clinical significance. In: Graham MD, House WF (eds) Disorders of the facial nerve. Raven Press, New York, pp 57–65

Stennert E, Böschen C, Gunkel A, Goebel HH (1994) Effects of electrostimulation therapy: enzyme-histological and myometric changes in the denervated musculature. Eur Arch Oto-Rhino-Laryngol Suppl Proceedings of the Seventh International Symposium on the Facial Nerve, Cologne, June 9-14. 1992, pp. S37–S41

Streppel M, Angelov DN, Guntinas-Lichius O, Hilgers RD, Rosenblatt JD, Stennert E, Neiss WF (1998) Slow axonal regrowth but extreme hyperinnervation of target muscle after suture of the facial nerve in aged rats. Neurobiol Aging 19:83–88

Streppel M, Azzolin N, Dohm S, Guntinas-Lichius O, Haas C, Grothe C, Neiss WF, Angelov DN (2002) Focal application of neutralizing antibodies to soluble neurotrophic faactors reduces collateral axonal branching after peripheral nerve lesion. Eur J Neurosci 15:1327–1342

Sulaiman OA, Midha R, Munro CA, Matsuyama T, Al-Majed A, Gordon T (2002) Chronic Schwann cell denervation and the presence of a sensory nerve reduce motor axonal regeneration. Exp Neurol 176:342–354

Sumner AJ (1990) Aberrant reinnervation. Muscle Nerve 13:801–803

Sumner BE, Watson WE (1971) Retraction and expansion of the dendritic tree of motor neurones of adult rats induced in vivo. Nature 233:273–275

Sunderland S (1950) Capacity of reinnervated muscles to function efficiently after prolonged denervation. Arch Neurol Psychiat 64:755–771

Suter DM, Forscher P (2000) Substrate-cytoskeletal coupling as a mechanism for the regulation of growth cone motility and guidance. J Neurobiol 44:97–113

Svensson M, Aldskogius H (1993) Synaptic density of axotomized hypoglossal motorneurons following pharmacological blockade of the microglial cell proliferation. Exp Neurol 120:123–131

Svensson P, Romaniello A, Arendt-Nielsen L, Sessle BJ (2003) Plasticity in corticomotor control of the human tongue musculature induced by tongue task-training. Exp Brain Res 152:42–51

Svensson P, Romaniello A, Wang K, Arendt-Nielsen L, Sessle BJ (2006) One hour of tongue-task training is associated with plasticity in corticomotor control of the human tongue musculature. Exp Brain Res 173:165–173

Szebenyi G, Gallaway JL, Dent EW, Kalil K (1998) Interstitial branches develop from active regions of the axon demarcated by the primary growth cone during pausing bahaviors. J Neurosci 18:7930–7940

Tam SL, Gordon T (2003) Mechanisms controlling axonal sprouting at the neuromuscular junction. J Neurocytol 32:961–974

Tam SL, Archibald V, Tyreman N, Jassar B, Gordon T (2001) Increased neuromuscular activity reduces sprouting in partially denervated muscles. J Neurosci 21:655–663

Tanaka E, Ho T, Kirschner MW (1995) The role of microtubule dynamics in growth cone motility and axonal growth. J Cell Biol 128:139–155

Tanaka E, Kirschner MW (1991) Microtubule behaviour in growth cones of living neurons during axon elongation. J Cell Biol 115:345–363

Tanaka E, Kirschner MW (1995) The role of microtubules in growth cone turning at substrate boundaries. J Cell Biol 128:139–156

Tankere F, Bernat I, Vitte E, Lamas G, Soudant J, Maisonobe T, Bouche P, Fournier E, Willer JC (2003) The hypoglossal-facial anastomosis in man. A model for studying peripheral and central nervous system plasticity. Rev Med Liege 59(Suppl 1):91–103

Targan RS, Alon G, Kay SL (2000) Effect of log term electrical stimulation on motor recovery and improvement of clinical residuals in patients with unresolved facial nerve palsy. Otolaryngol Head Neck Surg 122:246–52

Taylor KS, Anastakis DJ, Davis KD (2009) Cutting your nerve changes your brain. Brain 132:3122–33

Terrell GS, Terzis JK (1994) An experimental model to study the blink reflex. J Reconstr Microsurg 10:175–183

Terzis JK (2005) Microsurgical reanimation of the orbicularis oculi. In: Beurskens CHG, van Gelder RS, Heymans PG, Manni JJ, Nicolai JPA (eds) The facial palsies. Complementary APPROACHES. Lemma Publishers, Utrecht, The Netherlands, pp 187–194

Terzis JK, Papakonstantinou KC (2000) The surgical treatment of brachial plexus injuries in adults. Plast Reconstr Surg 106:1097–1122

Tetzlaff W, Bisby MA (1989) Neurofilament elongation into regenerating facial nerve axons. Neuroscience 29:659–666

Tetzlaff W, Bisby MA, Kreutzberg GW (1988a) Changes in cytoskeletal proteins in the rat facial nucleus following axotomy. J Neurosci 8:3181–3189

Tetzlaff W, Graeber MB, Bisby MA, Kreutzberg GW (1988b) Increased glial fibrillary acidic protein synthesis in astrocytes during retrograde reaction of the rat facial nucleus. Glia 1:90–5

Tetzlaff W, Alexander SW, Miller FD, Bisby MA (1991) Response of facial and rubrospinal neurons to axotomy:changes in mRNA expression for cytoskeletal proteins and GAP-43. J Neurosci 11:2528–2544

Tetzlaff W, Leonard C, Krekoski CA, Parhad IM, Bisby MA (1996) Reductions in motoneuronal neurofilament synthesis by successive axotomies: a possible explanation for the conditioning lesion effect on axon regeneration. Exp Neurol 139:95–106

Thanos PK, Terzis JK (1995) Motor endplate analysis of the denervated and reinnervated orbicularis oculi muscle in the rat. J Reconstr Microsurg 11:423–428

Thomander L (1984) Reorganization of the facial motor nucleus after peripheral nerve regeneration. An HRP study in the rat. Acta Otolaryngol (Stockh) 97:619–626

Thonnard J, Saels P, Van den Bergh P, Lejeune T (1999) Effects of chronic median nerve compression at the wrist on sensation and manual skills. Exp Brain Res 128:61–4

Tilney LG, Bryan J, Bush DJ, Fujiwara K, Mooseker MS, Murphy DB (1973) Microtubules: evidence for 13 protofilaments. J Cell Biol 59:267–275

Tiraihi T, Rezaie MJ (2004) Synaptic lesions and synaptophysin distribution change in spinal motoneurons at early stages following sciatic nerve transection in neonatal rats. Brain Res Dev Brain Res 148:97–103

Titmus MJ, Faber DS (1990) Axotomy-induced alterations in the electrophysiological characteristics of neurons. Prog Neurobiol 35:1–51

Tokiwa MA, Gaspar EM, Doering LC (1994) CNTF is superior to NGF as a survival enhancement factor for adrenal medulla cells in vitro. Neuroreport 5:549–552

Tolu E, Caria MA, Pugliatti M (1993) Responses of hypoglossal motoneurons to mechanical stimulation of the teeth in rats. Arch Ital Biol 131:191–200

Tomov TL, Guntinas-Lichius O, Grosheva M, Streppel M, Schraermeyer U, Neiss WF, Angelov DN (2002) An example of neural plasticity evoked by putative behavioral demand and early use of vibrissal hairs after facial nerve transection. Exp Neurol 178:207–218

Tonra JR, Curtis R, Wong V, Cliffer KD, Park JS, Timmes A, Nguyen T, Lindsay RM, Acheson A, DiStefano PS (1998) Axotomy upregulates the anterograde transport and expression of brain-derived neurotrophic factor by sensory neurons. J Neurosci 18:4374–4383

Trachtenberg JT, Thompson WJ (1996) Schwann cell apoptosis at developing neuromuscular junctions is regulated by glial growth factor. Nature 379:174–177

Travers JB, Norgren R (1983) Afferent projections to the oral motor nuclei in the rat. J Comp Neurol 220:280–298

Trojan DA, Gendron D, Cashman NR (1991) Electrophysiology and electrodiagnosis of the post-polio motor unit. Orthopedics 14:1353–1361

Tsuboya H, Tani T, Ishida K, Ushida T, Taniguchi S, Kimura J (2007) Quantitative sensory testing of cold and vibration perception during compression of median nerve at the wrist. Muscle Nerve 35:458–64

Tuttle GB, Mackey T, Steers WD (1994) NGF, bFGF and CNTF increase survival of major pelvic ganglion neurons cultured from the adult rat. Neurosci Lett 173:94–98

Uemura-Sumi M, Itoh M, Mizuno N (1988) The distribution of hypoglossal motoneurons in the dog, rabbit and rat. Anat Embryol 177:389–394

Umemiya A, Araki I, Kuno M (1993) Electrophysiological properties of axotomized facial motoneurones that are destined to die in neonatal rats. J Physiol 462:661–78

Valero-Cabre A, Navarro X (2002) Changes in crossed spinal reflexes after peripheral nerve injury and repair. J Neurophysiology 87:1763–1771

Valero-Cabre A, Tsironis K, Skouras E, Navarro X, Neiss WF (2004) Peripheral and spinal motor reorganization after nerve injury and repair. J Neurotrauma 21:95–108

Valls-Sole J, Tolosa ES (1989) Blink reflex excitability cycle in hemifacial spasm. Neurology 39:1061–1066

Valls-Sole J, Tolosa ES, Pujol M (1992) Myokymic discharges and enhanced facial nerve reflex responses after recovery from idiopathic facial palsy. Muscle Nerve 15:37–42

Van den Noven S, Wallace N, Muccio D, Turtz A, Pinter MJ (1993) Adult spinal motoneurons remain viable despite prolonged absence of functional synaptic contact with muscle. Exp Neurol 123:147–156

Van Praag H, Kempermann G, Gage FH (2000) Neural consequences of enriched environment. Nature Rev Neurosci 1:191–198

Van Swearingen JM, Brach JS (2003) Changes in facial movement and synkinesis with facial neuromuscular reeducation. Plast Reconstr Surg 111:2370–2375

Vaughan ED, Richardson D (1993) Facial nerve reconstruction following ablative parotid surgery. Br J Oral Maxillofac Surg 31:274–280

Vega QC, Worby CA, Lechner MS, Dixon JE, Dressler GR (1996) GDNF activates the receptor tyrosine kinase RET and promotes kidney morphogenesis. Proc Natl Acad Sci USA 93: 10657–10661

Veinante P, Deschenes M (2003) Single-cell study of motor cortex projections to the barrel field in rats. J Comp Neurol 464:98–103

Vetter P, Roth A, Hausser M (2001) Propagation of action potentials in dendrites depends on dendritic morphology. J Neurophysiol 85:926–937

Vignjevic D, Yarar D, Welch MD, Peloquin J, Svitkina T, Borisy GG (2003) Formation of filopodia-like bundles in vitro from a dendritic network. J Cell Biol 160:951–962

Vivó M, Puigdemasa A, Casals L, Asensio E, Udina E, Navarro X (2008) Immediate electrical stimulation enhances regeneration and reinnervation and modulates spinal plastic changes after sciatic nerve injury and repair. Exp Neurol 211:180–93

Watson WE (1965) An autoradiographic study of the incorporation of nucleic-acid precursors by neurons and glia during nerve regeneration. J Physiol (Lond) 180:741–753

Waxman B (1984) Electrotherapy for treatment of facial nerve paralysis (Bell's Palsy). Health Technology Assessment Reports, National Center for Health Services Research 3, 27

Welker WI (1964) Analysis of sniffing of the albino rat. Behaviour 22:223–244

Welt C, Abbs JH (1990) Musculotopic organization of the facial motor nucleus in Macaca fascicularis: a morphometric and retrograde tracing study with cholera toxin B-HRP. J Comp Neurol 291:621–636

Whitehead J, Keller-Peck C, Kucera J, Tourtellotte WG (2005) Glial cell-line derived neurotrophic factor-dependent fusimotor neuron survival during development. Mech Dev 122:27–41

Wiedemann E, Eggert C, Illert M, Stock W, Wihelm K (1997) Functional electromyography analysis of radial replacement operation. Orthopaede 26:673–683

Wiedenmann B, Franke WW (1985) Identificatiion and localization of synaptophysin, an integral membrane glycoprotein of Mr 38.000 characteristic of presynaptic vesicles. Cell 41:1017–1028

Williams HB (1996) A clinical pilot study to assess functional return following muscle stimulation after nerve injury and repair in the upper extremity using a completely implantable electrical system. Microsurgery 17:597–605

Williamson T, Gordon-Weeks PR, Schachner M, Taylor J (1996) Microtubule reorganization is obligatory for growth cone turning. Proc Natl Acad Sci USA 93:15221–15226

Wilson JR, Sumner AJ, Eichelman J (1994) Aberrant reinnervationfollowing hypoglossal nerve damage. Muscle Nerve 17:931–935

Wineski LE (1985) Facial morphology and vibrissal movement in the golden hamster. J Morphol 183:199–217

Yamada KM, Spooner BS, Wessells NK (1970) Axon growth: roles of microfilaments and microtubules. Proc Natl Acad Sci USA 66:1206–1212

Yamada KM, Spooner BS, Wessells NK (1971) Ultrastructure and function of growth cones and axons of cultured nerve cells. J Cell Biol 49:614–635

Yu W, Baas PW (1994) Changes in microtubule number and length during during axon differentiation. J Neurosci 14:2818–2829

Yu W, Baas PW (1995) The growth of the axon tip is not dependent upon net microtubule assembly and its distal tip. J Neurosci 15:6827–6833

Yu WH, Yu MC (1983) Acceleration of the regeneration of the crushed hypoglossal nerve by testosterone. Exp Neurol 80:349–60

Yu W, Ahmad DJ, Baas PW (1994) Microtubule fragmentation and partitioning in the axon during collateral branch formation. J Neurosci 14:5872–5884

Z'Graggen WJ, Metz GA, Kartje GL, Thallmair M, Schwab ME (1998) Functional recovery and enhanced corticofugal plasticity after unilateral pyramidal tract lesion and blockade of myelin-associated neurite growth inhibitors in adult rats. J Neurosci 18:4744–4757

Zerari-Mailly F, Pinganaud G, Dauvergne C, Buisseret P, Buisseret-Delmas C (2001) Trigemino-reticulo-facial and trigemino-reticulo-hypoglossal pathways in the rat. J Comp Neurol 429:80–93

Zhao C, Veltri K, Bain JR, Fahnenstock M (2004) NGF, BDNF, NT-3, and GDNF mRNA expression in rat skeletal muscle following denervation and sensory protection. J Neurotrauma 21:1468–1478

Zheng JL, Helbig C, Gao W-Q (1997) Induction of cell proliferation by fibroblast and insulin-like growth factors in pure rat inner ear epithelial cell cultures. J Neurosci 17:216–226

Ziv NE, Spira ME (1997) Localized and transient elevations of intracellular Ca^{2+} induce the dedifferentiation of axonal segments into growth cones. J Neurosci 17:3568–3579

Index